CW00705359

THE ENGINEERING REVOLUTION

THE ENGINEERING REVOLUTION

How the Modern World was Changed by Technology

Edited by Angus Buchanan

With contributions from:
David Ashford, Mike Bone, Angus Buchanan, Brenda Buchanan, Keith Falconer, Richard Harvey, Stephen K Jones, Robin Morris, Giles Richardson, Owen Ward

PEN & SWORD TRANSPORT

AN IMPRINT OF PEN & SWORD BOOKS LTD.
YORKSHIRE – PHILADELPHIA

First published in Great Britain in 2018 by
PEN & SWORD TRANSPORT
An imprint of
Pen & Sword Books Ltd
Yorkshire - Philadelphia

ISBN 978 1 47389 908 7

A CIP catalogue record for this book is available from the British Library.

Typeset in 10.5/13.5 pt Palatino
Typeset by Aura Technology and Software Services, India
Printed and bound in India by Replika Press Pvt. Ltd.

Pen & Sword Books Ltd incorporates the Imprints of Pen & Sword Books Archaeology, Atlas, Aviation, Battleground, Discovery, Family History, History, Maritime, Military, Naval, Politics, Railways, Select, Transport, True Crime, Fiction, Frontline Books, Leo Cooper, Praetorian Press, Seaforth Publishing, Wharncliffe and White Owl.

For a complete list of Pen & Sword titles please contact

PEN & SWORD BOOKS LIMITED
47 Church Street, Barnsley, South Yorkshire, S70 2AS, England
E-mail: enquiries@pen-and-sword.co.uk
Website: www.pen-and-sword.co.uk

Or
PEN AND SWORD BOOKS
1950 Lawrence Rd, Havertown, PA 19083, USA
E-mail: Uspen-and-sword@casematepublishers.com
Website: www.penandswordbooks.com

Contents

Preface

This book is the work of current members of the History of Technology Research Unit (HOTRU) at the University of Bath. The Unit was established as the Centre for the History of Technology at the Bristol College of Science and Technology in 1964, two years before it became the University of Bath, and in over fifty years the Unit has achieved some modest successes in advancing knowledge in several areas of the history of technology, including the history of steam technology, the development of naval power, the history of gunpowder, the emergence of the engineering profession, industrial archaeology, and the conservation of industrial artefacts. The authors of this project on *The Engineering Revolution* are all members of the HOTRU Seminar, as Visiting Fellows, Rolt Fellows, or local supporters. Their individual authorship of chapters is attributed at the beginning of each piece. The Editor is grateful for the assistance and patience of his colleagues in helping him to compile this composite text.

The history of technology is an account of its tremendous transformative effect on human societies in the last two million years, but particularly the last three centuries – the period of 'High Technology'. This emphasis on the recent past is justifiable because it can be regarded as a 'Revolution' in terms of physical progress, there being no reasonable doubt that humankind has acquired huge powers in the period since 1700, compared with the relatively slow progress of previous centuries. The spectacular achievement of the period, however, has also raised formidable new problems, many of which remain to be resolved, and so confront the modern world with novel challenges. The greatest of these is the task of surviving in a world made dangerously unstable by the technological power that is now available to human societies that are only learning slowly how to control it. The challenge is thus to understand these technological processes and to learn how to direct them towards creative roles before they are misused to destroy our civilisation.

The standpoint of the authors is unavoidably British and European, although we attempt to set this view of the technological revolution in a global context and recognise the worldwide implications of the subject. Without the expertise, however, to undertake a more comprehensive analysis of its development in large parts of the world, we hope that we are able to present a sound and satisfactory overview of the scope of the subject. That at least is the intention of the authors in providing an introduction for anybody who is interested in exploring the history of technology. We are thinking in particular of intelligent young people preparing themselves for university admission, but we hope that it will find favour with general readers also.

Unfortunately delays in preparation and publication have caused us to miss the fiftieth anniversary celebrations at the University of Bath in 2016, but we hope that the book will nevertheless be accepted as our contribution to these celebrations. As a research centre that has been with it since its outset, we congratulate the University on its outstanding academic achievements and dedicate this project to it as a token of our thanks for the support of the University over all these years. We also express the hope that this book will help to make the case for the history of technology in the future teaching and research of the University.

Angus Buchanan, Editor June 2018

Dedication

To the Students and Staff, Past and Present, of
The University of Bath, in recognition of fifty years
of academic achievement 1966–2016.

CHAPTER 1

Introduction – Technology in Context

Angus Buchanan

Technology is the study of the way human beings make artefacts and do things with them. The ability to acquire such techniques, other than by instinct such as that by which birds build nests, is virtually unique to human beings on Planet Earth. Indeed, the possession of such power is one of the chief qualities distinguishing the human species, which makes 'technology' at least two million years old. For most of this huge period of time, the development of technical skills has been extremely slow and repetitive. While the climate changed with a succession of prolonged Ice Ages, and while the level of the oceans rose and

Fig 1.1: Stonehenge, 4,000 years old, marks the transition from Low to Median Technology. Its builders not only moved and raised the stones but worked them into shape and fitted them together. (*Angus Buchanan*)

fell, 'hominids' – man-like species – struggled to survive in the sub-tropical region of central Africa where they first emerged from the jungle. Then, as the climate became warmer with the end of the last Ice Age about 20,000 years ago, our own species – *Homo sapiens* – emerged from a long period of competition with rival hominids and undertook the tremendous task of taking control of their environment, although they were not able to see it in such grandiose terms.

They tamed animals to help them in their hunting, and improved the crude stone implements of their predecessors to create well-shaped tools and weapons. They devised intricate techniques of linguistic communication and discovered how to make and control fire. They expressed themselves in haunting cave paintings of animals that they hunted for food and for furs to provide their first clothes. We know very little about the lives of the many generations of these people, except what we can interpret from the archaeological fragments that they left behind. But we can be sure of their hardiness and their determination to survive and pass on their bloodline to their children, because without such qualities we would not be here to wonder at their achievements in a largely hostile environment.

Moreover, they spread as their population increased and they hived off communities of hunters and their families who moved away from the African heartlands in search of food. Some moved north towards milder climates and new environments, to which they adjusted themselves and journeyed on over many generations; some eastwards, to populate South East Asia; and others westwards, around the Mediterranean and, at length, into Europe. Some, such as the people who moved into Australasia about 40,000 years ago, became isolated by rising sea levels, and established a virtually static existence for many millennia. The Americas were populated some time later, as tribes crossed the Aleutian chain of islands to Alaska and then moved southwards. It is possible that others managed to cross the Atlantic Ocean from the east on primitive rafts, but that remains a matter of speculation. Others, in India and China, made relatively rapid technological progress. For all parts of the land surface of the world that they managed to inhabit, however, the struggle for survival remained arduous, absorbing the energy of individuals and limiting their aspirations.

Phase One: Low Technology

Gradually, human societies increased their power – their ability to make and do things – by improving their tools and weapons, and thus began a long process of technological development. Although new ideas must have come initially from individual people, we have no way of knowing anything about such individuals, but we can recognise their collective achievements through their surviving artefacts. Anything they made of wood or vegetable fibre or animal skins has usually perished, but the survival of stone implements provides ample evidence of advancing technical skills. In the earliest human settlements, natural stones were adapted for use in hammering and cutting. In later settlements, the evidence of stone chippings demonstrates that techniques of shaping stone to produce more efficient tools and weapons had been acquired, presumably by striking one stone with a harder stone, or by subjecting it to a laborious exercise of grinding one stone against another. By the end of the Old Stone Age, some 10,000 years ago, the remains of well-shaped axes and spear-heads demonstrate great skill in stone working, and frequently bear evidence of shafting on to handles or poles which have long-since disappeared. By this time, also, the more advanced groups of human beings had discovered the control of fire, which enabled them to prepare more varied and nutritious diets and to keep warm and safe at night. Thus equipped, they were able to survive in variable climates, and so to undertake more ambitious migrations than any they had attempted before.

All this can be confidently based upon the surviving archaeological evidence, although this tells us nothing about individuals and little about the social organisation of these early groups of human beings, or of what they believed or how they mastered sophisticated means of linguistic communication. There is not much to show for this long period of some two million years, except the vital fact that human beings mastered the basic techniques of survival in a variety of different conditions and climates. And if we pursue the notion that technology is about the power to make and do things, these achievements were all made with the power available from human hands and feet, making this the defining quality of the first phase of the history of technology, the period dominated by 'Low Technology'. Later phases achieved much more in a much shorter period of time, but they could not have done so without the two million years of trial and error, by which human beings who must for ever remain anonymous to us, established the basic techniques for survival in a challenging environment.

On this basis, we can distinguish two subsequent phases of technological evolution: that in which techniques were acquired for using the power of animals and of naturally replenishable sources of power such as the wind and falling water ('Median Technology'); and that in which the power of heat engines was harnessed to convert the energy of burning fuel into novel ways of making and doing things on a scale that could not previously have been imagined ('High Technology'). Median Technology dominated the period from about 2000BC to AD1700, and High Technology has flourished in the period since 1700. We will look briefly at the way in which technology transformed society in these two periods.

Phase Two: Median Technology

Stonehenge was built in stages over many years, but it was certainly in the course of construction and use around 2000BC. It is thus not unreasonable to take this outstanding monument as marking the transition from the Low Technology of the archaeological Old Stone Age to the Median Technology of the New Stone Age, and the use of metals in the large and well-organised societies that succeeded it. Earlier societies had moved large boulders, forming them into circles and long causeways, as at Avebury in Wiltshire, Callanish in Scotland, and Carnac in Brittany. But at Stonehenge the stones were not only placed in a circle: they were shaped and capped with similar stones which were fixed in place by mortise and tenon joints, representing a dramatic advance in stone working techniques and engineering skills. We can still only speculate about how it was achieved and to what purposes it was devoted, but while we can surmise that it was intended for some sort of religious or communal use, we can be certain that its construction involved a large number of highly skilled craftsmen and an army of labourers and supporters. Such a community implies an astonishingly high level of social organisation amongst people who were far more numerous than the small tribal societies that had previously been the norm in Western Europe, although large civilised societies had already been established in Egypt, Sumeria, India, and China. Even though Stonehenge preserves the anonymity of its builders, it can be seen as being on the cusp of the transition to a new phase of Median Technology in Western Europe.

The dynamic force of this transition probably came from an increase in population that had stimulated a change from an existence based upon hunting and collecting food to one that managed its own food supply by adopting agricultural techniques of growing crops and keeping animals for food, both as meat and for their milk. This implied a transition from a nomadic society, for ever on the move in search of better hunting grounds, to a more settled society, in which people cleared the land of trees and undergrowth in order to cultivate crops for food to sustain themselves and their herds of animals. In such settled societies, new forms of specialisation arose: men who had previously been hunters undertook the heavy duties of tilling the soil, and women who had previously been food gatherers became cooks, processing food over fires and in ovens.

Further specialisation followed, with a few families emerging as millers to grind the corn and other grains, and some as potters to make culinary vessels of all sorts, acquiring in the process knowledge of how to spin the soft clay they were working on a wheel by operating a crank with their feet. Others, again, learnt skills in brewing beer and distilling spirits, and in making clothes and shoes. The net result of such activities over a substantial area was a significant rise in productivity, a further growth of population, the rise of towns for the exchange of agricultural produce and as a means of social control and defence, generating further specialisation in luxury goods and refined metal working. In a word, within a short space of time in comparison with previous technological developments, civilisation emerged.

Civilisation did not occur uniformly, but in favoured temperate latitudes where large rivers watered fertile land which could be readily adopted for agriculture and for the development of towns. Such towns were the characteristic feature of civilised communities, even though most of the population long remained committed to providing food by agricultural and pastoral pursuits, and satisfactory town life depended on techniques that had been virtually non-existent before. These included the possession of skills in measurement, essential for the precise definition of property ownership; a mode of recording commercial transactions between individuals and parties, so that trade could operate smoothly at a distance and over time; and the creation of a monetary system accessible to the whole population. All these requirements assumed the existence of skills in literacy and numeracy, which became the crucial attributes of the new social specialisation of a class of clerks who could undertake measurements and keep records. These clerks frequently became clergy, controlling the important religious functions of the community, such as maintaining and interpreting the scriptures and traditions, and determining the feasts and the public commemorations. In China, they became the powerful class of 'mandarins', a non-hereditary governing class which served the ruling monarchy and to which access was by special training and examination.

In other civilisations the role of the clerks was rather less conspicuous, but always it was essential to the smooth running and efficiency of the society. As well as clergy directing religious affairs, other important specialisations, in administration, science, and historiography, were generated. The administrative activities of the clerks in running the government, collecting the taxes, and maintaining law and order, were legion and need not concern us in detail. In science, however, the clerks made the systematic study of their environment possible by their accumulation of records concerning the movement of the stars and planets, the relative productivity of different provinces, the nature of their mineral resources, and many other subjects. The use to which they put this information was at first astrological rather than astronomical, but came to provide a basis for valuable scientific speculations which were properly astronomical. Similarly, the information about the nature and whereabouts of materials had at first an alchemical value rather than one representing chemical or biological science, but it gave a realistic starting point for such studies. It is not unreasonable, therefore, to see science beginning with the emergence of literacy and numeracy in these first civilisations.

The same line of thought leads to the realisation that history – the systematic study of the past – is also a product of the early civilisations, whose clerks kept accounts of previous kings and other rulers, frequently going back many generations. While granting that the earliest parts of such genealogies were almost inevitably legendary or even mythological, they remained a valuable source of information for clerical scholars to examine and analyse, thus comprising genuinely historical studies. The fact that archaeologists have very recently – in the last two hundred years – developed ways of tracing the record of human beings back many millennia through their physical remains and artefacts gives a new dimension to the study of history, but it does not alter the supposition that the realistic study of the past only began with the first civilisations.

The observation that the history of technology began two million years ago, while the study of science and history only began some 4,000 years ago with the first civilisations, makes it all the more remarkable that technology has received such poor attention from clerks and scholars. The trouble was probably that the considerable achievements of technology by the time of Median Technology in which science and history were born, were already being taken for granted as having been present from the beginning. The foreshortened view of the past shared by the early clerks thus meant their own tools, weapons and techniques were seen as part of the established order and so not worth serious study. These artefacts, moreover, were usually the product of artisans and slaves, who were already regarded as the bottom of the social scale, and as such not worth the attention of clerks. This certainly contributed to the strange lack of interest in technology shown by these early scholars, and their disinclination to deal with it persisted in subsequent generations. As far as science is concerned, it might have been expected that it would have received as little attention as technology, as the two disciplines share a great deal of common ground and have been mutually supportive in many respects, but as science involved so much speculative thought it was more attractive to the early clerks than the more practical concerns of technology. Nevertheless, the processes of speculation and invention fed into each other, as when improvements in tools and instruments such as telescopes and microscopes became of incalculable value in science, and science in return has stimulated many technological initiatives. But this inter-relationship only became apparent much later, in the period of High Technology, and until then the two fields of study tended to keep to their own special interests.

There were, however, significant technological advances in the early civilisations, which deserve to be taken into account and to justify a distinction between an earlier stage of Median Technology, during which reliance for power was on human and animal strength, directed primarily towards land clearance and drainage for systematic agriculture; and a later stage, during which experiments in harnessing 'natural' sources of power such as that of wind and falling water, gradually increased productivity and competence. The new human attention to agriculture, in the first place, underwent considerable development, as various grains and other crops were tried and improved by careful selection over the seasons. Animal stock was also improved by selective breeding,

Fig 1.2: Primitive ploughing. Thirty years ago, traditional farming methods still prevailed in parts of China such as these fields near Wuhan where, in intensively farmed land, a water buffalo pulls a simple plough. (*Angus Buchanan*)

and tools were developed, such as the plough, which, when drawn by oxen, provided farmers with a tremendous help in breaking and turning the soil. This was made even more efficient when the use of horse shoes and collars made it possible to adapt the horse to drawing the plough, and new ways of maintaining soil fertility were explored by manuring and keeping one field (out of two, and subsequently out of three) fallow in rotation – all contributing to a gradual increase in productivity. Cooking techniques were advanced by building ovens adjacent to fires so that the heating could be partially controlled. Strong beverages were brewed, and stronger ones became attainable with the technique of distilling in order to concentrate the alcoholic content. Cooking and drinking vessels of many shapes and sizes were produced by the new skills of the potter at his wheel; and the wheel itself, probably a Neolithic invention in Asia, transformed the problem of transporting heavy loads in the shape of the wheeled wagon.

The emergence of civilisation with the growth of life in towns brought an acceleration in the processes of technological innovation

Fig 1.3: Laxey Waterwheel. The 'Lady Isabella' waterwheel at Laxey on the Isle of Man is the largest surviving in Britain. It drove pumps to drain a lead mine and has been preserved in working order. (*Angus Buchanan*)

to serve the requirements of urban communities. The provision of water for consumption and cleansing became an urgent necessity, leading to the diversion of rivers and the construction of aqueducts. Building in all respects, using timber, stone, and mud-dried or fired bricks to construct houses, temples, and monuments, underwent tremendous expansion, but this was regarded as architecture and received more attention than mere technology. Over the centuries immediately before and after the beginning of the Christian era, however, there were exciting innovations in water-wheels and windmills, in sailing ships and the use of the compass, which together encouraged extensive ocean voyaging and led to the discovery of new trade routes and new continents. At the same time, the trains of gearing used in mill work opened the way to intricate mechanical devices such as the clock; and the discovery of gunpowder brought a long-running military revolution, as nations equipped themselves with cannons and other guns that underwent continuous improvement in fire-power and efficiency.

Metals generally, beginning with the softer ones such as copper and gold which were the easiest to work, but going on to tin which, when fused with copper in the right proportions, produced bronze; and then iron, which required a particularly high temperature to reduce it from its ore, but was sufficiently hard to make a wide variety of tools and weapons, were worked on an increasing scale and used in multifarious ways. And finally, as the phase of Median Technology drew to a close, the invention of the printing press and of paper produced from rags initiated one of the great transformations of Western Civilisation with the Renaissance, the Reformation, and the Scientific Revolution.

Fig 1.4: Cranbrook Windmill. At Cranbrook in Kent this windmill dominates the village street. It is known as a 'smock mill' because the flared wooden tower resembles the apron of a miller. (*Angus Buchanan*)

It is certain that much was achieved in the phase of Median Technology, although sufficiently slowly to be assimilated without too drastic an upheaval – except possibly in the case of gunpowder and the printing press – to impinge on the public mind. Throughout it the more advanced societies acquired techniques that allowed them to harness the wind and water-power to supplement that of human and animal labour. The wind was probably the first such 'natural' source of power to be harnessed, with primitive sails propelling boats, but water-wheels were certainly prominent in the Roman Empire, in both horizontal and vertical configurations, making a valuable contribution to the arduous task of grinding grain into flour. Windmills were introduced to Europe, probably from the Middle East or even China, in the twelfth century AD. Throughout Western Europe, productivity increased and the population gradually grew, except when interrupted briefly by epidemic diseases or catastrophic warfare. Knowledge of the world was greatly widened by the Western voyages of discovery and the waves of commercial exploitation, missionary endeavour, and the imperial aggrandisement that followed it. Technology contributed substantially to this development, opening up possibilities and dangers that materialised in the phase of High Technology.

Phase Three: High Technology

High Technology began with the introduction of the steam engine in 1712. For once, we can give a particular date as being significant in this chronology, because it marks the operation of the first engine using steam power effectively, representing an important shift of power from natural and renewable sources to heat engines using the energy from burning fuels to drive a machine. In 1712 the first steam engine, designed by Thomas Newcomen, began to work at a coal mine in Dudley in the British Midlands, inaugurating a century and a half of rapid technological and industrial development that constituted the first stage of High Technology. This period was followed by a second stage, beginning in the 1880s, of even more rapid technological transition in which steam engines were largely replaced by other prime movers.

Newcomen's engine was, strictly, an atmospheric engine, because the steam was condensed in an open-topped vertical cylinder to produce a partial vacuum, causing the pressure of the atmosphere on the upper face of a piston in the cylinder to drive the piston down the cylinder. The upper end of the piston rod was attached to

Fig 2.5: Eddystone Rock Lighthouse. John Smeaton erected the first modern-style lighthouse in 1759. When replaced at the end of the nineteenth century, the upper two-thirds were dismantled and re-erected on Plymouth Hoe. (*Angus Buchanan*)

a swinging beam, so that the depression of the piston at one end of the beam caused pumping rods at the other end to rise in the shaft of the mine, drawing water with them and thus beginning to pump dry the working tunnels of the mine. Releasing the vacuum in the cylinder then caused the weight of the pump rods to pull the piston back to the top of the cylinder, and this reciprocating stroke could be repeated indefinitely.

This ingenious arrangement was the germ of all subsequent heat engines. Thomas Newcomen, an iron merchant from Dartmouth in Devon, is regarded as the inventor of the Dudley engine, and this is justified even though the component parts such as the effect of condensing steam in a cylinder, the movement of a piston, and the use of a swinging beam to transfer the power stroke, had already been investigated, and Newcomen's genius was that of bringing them together in a viable machine. It was a bulky and inefficient machine, requiring a large supply of fuel, but at a working colliery with plenty of small coal available at the pithead, this was not a serious problem, and so the invention was rapidly adopted in the coal mining areas of Britain.

It also spread quickly to France, Austria, Sweden, and the British colonies in America. It remained expensive and inefficient and confined to simple pumping functions, however, until James Watt, a Scottish instrument-maker, introduced his improvements in partnership with Matthew Boulton, a Birmingham entrepreneur, in the last quarter of the eighteenth century. Watt's separate condenser, parallel motion, rotary action, centrifugal governor, and double-action, did not make the steam engine less expensive, but they did make it spectacularly more efficient and more useful, because with rotary action it became possible to harness steam power directly to a vast array of factory machines. When the Boulton & Watt patent expired in 1800, they had sold several hundred engines, most of which were still at work in mines and factories all over the world. It was clear to perceptive observers that the steam engine had ushered in a new age of technology.

This was only the beginning of the steam revolution. Watt had always resisted using high pressure steam in his engines, but when he retired in 1800 some very talented inventors, such as Richard Trevithick, a Cornish mines captain, who had devised boilers capable of raising steam safely at much higher pressures than Watt had permitted, applied it to steam engines. When used in a carefully determined cycle, this proved to be a very economical machine, which became known as the 'Cornish engine' and was used all over the world where deep mining for metals required such machines. Trevithick also pioneered the use of his engine in a locomotive form, although he was discouraged by its early difficulties so that the full development of the steam locomotive was left to George Stephenson and his son Robert and a host of other talented engineers. In the process, it was tried in many different forms and adapted for marine propulsion and played a formative role in the nineteenth century growth and transformation of industry and transport systems.

However impressive, the dominance of the steam engine was short-lived. By the middle of the nineteenth century, it began to be challenged by two other sources of power – the internal combustion engine and electricity. The internal combustion engine is a heat engine like the steam engine, from which it derives much of its working form, but whereas the steam engine burns its fuel externally to generate steam, its working fluid, the internal combustion engine burns it internally, the fuel – in liquid or gaseous form – being injected into the working cylinder to be ignited – usually by an electric spark – and so to move the piston. The first successful engines of this type used coal gas as fuel, but this was soon displaced by oil fuels, particularly petrol or gasoline, which had the enormous advantage of being easily transportable and so ideal for locomotive purposes. By the 1880s, such engines were being installed in vehicles to become the first automobiles or motor cars, and they soon began to replace steam engines in many other industrial and transport applications. In addition, the first decade of the twentieth century witnessed the invention of the aeroplane powered by oil-fuelled engines, enabling another vast transport revolution to take place.

But the internal combustion engine never enjoyed the near-monopoly of power sources such as that of the steam engine before it. For one thing, the steam engine continued to make improvements, and by adopting the turbine form, in which the steam acts like the water in a water turbine to drive a wheel either by passing through vanes on it or by striking buckets on its rim, a series of inventors devised a machine which produced rotary motion by direct action and was ideal for the generation of electricity. Although not itself a 'prime mover' in the sense of being an initial source of power, electricity produced by steam turbines does constitute such a system and provides a strong competitor to internal combustion in many transport and industrial applications. It also provides in abundance things such as light and heat and electronic capabilities that are not so readily available from any other source.

The development of this omnipresent power could not begin until eighteenth century scientists such as Benjamin Franklin had tried to understand the nature of lightning and the phenomena of magnetism and static electricity. But then in 1800 Volta demonstrated a way of producing an electric current with the device that became known as a 'voltaic pile', consisting of a stack of discs of different metals reacting with each other in a chemical solution. Michael Faraday was able to take this further in 1831 by showing that an electric current could be created in a coil of wire by rotating it between the poles of a magnet, and the way was then open to both the electric dynamo and the electric motor, although it took another forty years to develop these concepts into practical engineering possibilities. So it was that the widespread distribution of electricity from steam turbines in power stations was introduced in the 1880s, to play an enormous part in determining the achievements of the second stage of High Technology.

The acquisition of vast new sources of power that has characterised High Technology has had tremendous consequences for the modern world. In the first place, it has contributed to the unification of what was initially a Western European civilisation into a truly World or Global civilisation in the sense that its dominant features – easy access to power for manufacturing industry, civil engineering structures, transport, and communications – are now available in virtually every part of the inhabited world. It has thus made possible immediate contact between people in all parts of the world, with rapid transport by land, sea, and air, and rendered possible higher standards of living for everybody permitted to enjoy them by the particular society in which they live. On the other hand, it has brought into sharp rivalry societies with different perceptions and aspirations, and by putting at their disposal an armoury of horrific weapons it has presented the world with a constant threat of devastating conflict.

The scope of this ongoing technological revolution can be indicated by the rise of Information Technology. From comparatively slow beginnings in electric telegraphy in the 1830s, the pace of development quickened in the 1870s with the advent of the telephone, and hard on the heels of this popular innovation in human communication came the invention of electronic techniques such as the radio and then television. The first public television services were introduced just before the Second World War, and only became fully developed in the 1950s. By this time they had combined with the exploration of electronic computing machines, which had been pioneered during the 1939–1945 war, to produce an astonishing range of personal computing devices. These have made possible the Internet and the world-wide-web of instant communication all round the world – and beyond, in the triumphs of space exploration of the Moon and Mars and the rest of the solar

Fig 1.6: The NASA Space Shuttle coming in to land here was until recently the work-horse of the American Space Program. It was launched with rocket engines but landed with a controlled glide. (*NASA*)

system. In a very short space of time technology has made available instruments that can match human thought processes and put powers at the disposal of human societies that are of breath-taking immensity. Provided, that is, that they are wisely used and not abused.

The extraordinary speed with which the transformations of High Technology have been achieved may thus be considered a Technological Revolution, and it is this historical process of the last three centuries that is the subject of the present book. The concept of revolution requires some justification, because to engineers it retains its basic sense of turning round as with a wheel. But modern historians have become familiar with its use in a political context to describe a sudden change in social organisation such as forms of government, often accompanied by violence as in the French Revolution of 1789 or the Russian Revolution of 1917. They have also applied it to more general changes in the social order, as in the Agricultural Revolution and the Industrial Revolution, even though these have been less sudden events than political revolutions, and have been spread over many decades. In this context the period of High Technology, which has continued for the last three centuries and is still in process, can justifiably be said to constitute a 'revolution' in comparison with the much more protracted developments of the earlier phases of technological history, and so it is this sense that it is used here to determine the subject matter of this book. Without losing sight of the long roots of our narrative, therefore, we confidently describe the phase of High Technology as having the essential features of a Technological Revolution.

Further reading

Buchanan, Angus *The Power of the Machine,* (Viking, London, 1992).

Derry, T K and Williams, Trevor: *A Short History of Technology* (Oxford, 1962).

Mumford, Lewis: *Technics and Civilization* (Routledge, London, 1934).

Pacey, Arnold: *Technology and World Civilisation* (MIT Press, 1991).

Kranzberg, Melvin, and Pursell Jr, Carroll W (eds): *Technology in Western Civilization* (New York OUP, 1967).

Uhling, Robert (ed.): *James Dyson's History of Great Inventions* (Constable, London, 2001).

Usher, Abbott Payson: *A History of Mechanical Inventions* (Harvard, 1954).

White Jr, Lynn: *Medieval Technology and Social Change* (Oxford, 1962).

Singer, Charles et al (eds): *A History of Technology* (5 vols, Oxford 1954, plus 2 vols on the twentieth century) (ed. Williams, Trevor, Oxford, 1978).

Chapter 2

Feeding the People

Mike Bone

Food and drink are essential to human existence and their provision has been the major task of most of mankind for much of our past. In modern times, as the percentage of workers in advanced societies engaged in agriculture falls year by year, it is still one of the largest employment sectors in the world, second only to the service industries. It is therefore not surprising that the earliest tools were fashioned to find and process food and drink and that subsequent providers have been quick to make use of developments in science and technology in their attempts to sustain an ever-growing population.

Hunter-gatherers: food and drink in the age of low technology

The development of tools and techniques in this period has been covered in our introduction. However protracted and rudimentary these advances might seem, it has been estimated that a hunter-gathering economy could sustain a world population of twenty–thirty million people, about three times the estimated population at the time, albeit at a very low level of well-being. Recent research in anthropometric history has established the relationship between height and standards of living and, if we accept Louis Leaky's assumption that our earliest ancestors averaged only about 4ft 3in (1.3m) in height, comparison with today's situation vividly illustrates advances over the centuries.

The first agricultural revolution

Improvements in the techniques of hunter-gathering have, of course, continued to the present day, with the sophistication of the modern fishery as perhaps the best example. The spear, the hook, the line and the net were all in use during the Mesolithic period. However, the Neolithic or New Stone Age brought about a fundamental change in food production and the development of technology – the domestication of animals and the development of a settled pattern of tillage which marks the birth of agriculture as we know it. This 'Neolithic Revolution' as V Gordon Childe called it, ushered in the age of Median Technology. Agriculture would now produce a sufficient surplus of food that could support a much larger population, many of whom were no longer required to live in the country or work the land. This first agricultural revolution happened at different times on different continents. It is generally thought to have originated in the fertile crescent of the Middle East c.10000BC, but similar changes have been noted in South East Asia before this. The change had taken place in America by 7000BC.

The domestication of animals preceded the advent of tillage, perhaps starting when hunter-gatherers synchronised their movements with the seasonal migrations of animals such as goats and sheep. It seems that certain species such as dogs and cats first followed these migrating groups as scavengers prior to their domestication. The archaeological record suggests that they were followed by reindeer, goats, sheep and pigs. The advent of a more settled pattern of farming led to the keeping of cattle for meat, milk and haulage, bees for honey – the only sweetener before sugar became available – and the use of the camel, horse and ass as transport animals. Thus, by the second millennium BC most of the major species of farm animals or pets as we know them today had been domesticated and, by preventing contact with wild animals, our ancestors were able to improve the appearance and yields of their newly acquired flocks.

The cultivation of plants probably began by accident when migrating groups revisited previous camp sites and discovered crops that had grown from discarded or lost seed that had produced a higher yield on a small plot of ground than could be harvested in the wild. The development of tillage across the world was influenced by differences of climate, surface geology and culture but all such systems depended upon the basic techniques of expanding and maintaining the fertility of soils best suited to a chosen species. These crops (cereals, pulses, fruits and vegetables) have remained the same through the ages as have the stages of cultivation: turning the soil, planting seed, controlling weeds, watering and fertilising, harvesting and storage of the crop.

The thin soils of the Near East were easily turned by light ploughing. Tomb paintings in Egypt show an early type of plough worked by two men – one pulling and the other guiding the plough – which from c.3000BC had developed into one drawn by oxen, presumably to provide the greater depth of cultivation needed to maintain the fertility of the soil on continuously-cropped land. The plough was to remain the focus for future development in the age of Median Technology, with simple hand tools such as hoes and sickles employed in the later stages of the growth of the crop and its harvesting.

In the river valleys of the Near and Middle East it was essential to conserve and bring water to the growing crop. An illustration from a tomb at Thebes of c.1500BC shows a shaduf that, with the aid of a beam, raised water from the Nile that was emptied into an adjacent irrigation channel. These simple machines were later erected in series and augmented or replaced by bucket chains – these were used to water the Hanging Gardens of Babylon – and ox-driven wheels with buckets that worked with the aid of gearing. Water and sediment from the annual flooding of the Nile was stored and distributed to the fields by systems of tanks, sluices, aqueducts and canals. The flooding of the Tigris and Euphrates were less predictable than that of the Nile and required more complex systems of storage and distribution to deal with these irregular inundations. Management of water was also crucial to rice cultivation in Asia. In areas distant from the rivers, wells using a rope and tackle are in evidence from c.1500BC. Deeper artesian wells were also dug in Egypt whereby the pressure of the aquifers forced the water to the surface. Such developments were essential for agriculture but also necessary to provide water for developing cities.

The age of Greece and Rome brought little by way of major developments in tools and processes, but farming on the large, slave-worked estates of Rome was to mark the high-point of agricultural technology in the Mediterranean world during ancient times. Environmental conditions in Greece and Italy were similar, both having light soils, a climate characterised by alternate periods of drought and heavy rain and a shortage of land to keep large numbers of livestock for meat and manure. Consequently both used a light plough, made more effective by the addition of iron in Roman times, and maintained soil fertility by frequent ploughings, harrowing to remove weeds and to break up the ground, and alternate fallows. This careful husbandry, however, produced only a fourfold yield increase. The Romans also introduced the corn-drying kiln to prepare grain for milling. The skills and enterprise of the Romans in supplying water and its distribution via aqueducts is well known but this period also brought the beginnings of systematic drainage of some of the marshy river valleys of Italy as well as areas of their conquered territories.

Fig 2.1: Medieval Farming. Plough and Harrow from the Golf Book, Bruges c.1520. (*Credit: (c) The British Library: Additional Mss.24098 f.26v.*)

Many of the tools of these early civilisations – the sickle, the wooden hurdle, the ladder and single-handed sheep shears – were to remain in common use in post-Roman times, as can be seen in the illustrations in the Psalters and books of hours produced by the monks of medieval Europe.

In the later stages of the age of Median Technology our focus switches to northern Europe and the developments that were to prepare the way for a second agricultural revolution. Once again, the plough was at the forefront of innovation, with the addition of a mould-board to break up and turn heavy clay soils, many recently taken in from 'waste' and woodland. This happened in the eleventh century but it is thought that a similar improvement was introduced in China some 2,000 years before this. As in previous centuries, most of the other work in the fields was performed with traditional hand tools, with the additions of the flail (for threshing) in the eighth century and the scythe (for mowing) that came into common use some four centuries later.

New lands and new crops in the age of discovery

The development of ocean-going ships and techniques of navigation had extended the market for farm produce since ancient times, as evidenced by the relics of the wine and olive oil trading of the Phoenicians that have been discovered around the Mediterranean. The demand for grain to feed the mills of Rome also created a massive trade, some coming from as far away as the frontier province of Britain. It was, however, the pioneering voyages of John Cabot and Christopher Columbus in the last years of the fifteenth century that had a much greater impact on world agriculture, diet and consumption, with the discovery and transfer of 'new' crops across the globe. The potato, grown in the Andes long before the birth of Christ, was introduced into Europe by 1575 after Spanish conquests in South America. Its take-up took some time but it grew in popularity in France and England in the eighteenth century and was particularly popular in Ireland, with disastrous consequences when the crop failed in the mid-nineteenth century. Maize was not then considered suited to northern Europe, but was introduced to the East Indies and Africa by the Portuguese. Cane sugar and rice were taken from Europe to America, sugar having been grown in Spain and Sicily and traded extensively by the Venetians. By the mid-seventeenth century, large sugar plantations worked by slaves had been established by French and English settlers in the Caribbean and North America. The sugar was partially refined there and sent to Europe for the final stages of manufacture, where it had replaced honey as the sweetener of choice by 1700.

New beverages were also introduced into Europe. These included tea (first shipped from China by the Dutch in the early seventeenth century), coffee (originally from Africa but later available from Brazil and Java) and cocoa (from Africa) which gained popularity in the nineteenth century when sweetened with sugar. Alcoholic drinks have a much longer history. Vines were first grown in the Near East but viticulture and winemaking were mastered by Greece and Rome, with extensive exports across their respective empires. Whilst wine was the drink of choice in southern Europe, beer was more popular in the north. The fermentation of barley, as of grapes, probably happened first by chance and was replicated by those who enjoyed both the taste and after effects of this accidental brew. Brewing originally developed in the Near East, with no less than nineteen different beers being available to the Sumerians. In medieval Europe the major innovation in brewing was the boiling of the malt extract (or wort) with hops from the thirteenth century onwards. The hopped variety of the drink had a distinctive bitterness, kept for longer than its predecessor and was by far the safest beverage available to most until the improvement of urban water supplies in the later nineteenth century.

Distillation of spirits owes its origin to the alchemists of the first century AD, when the product was used medicinally. Spirits and liqueurs became popular drinks some 1,000 years later but excessive consumption had become a problem in some parts by 1300AD, as were the later varieties of gin, brandy and whisky at the end of the age of Median Technology. Another product of the Age of Discovery that began as a medicine was tobacco. The plant originated in South America, was taken to Europe and grown extensively in the sixteenth century, but became available in greater quantity and larger amounts after the seed was taken to North America in 1612.

Fig 2.2: A Large Brewhouse, 1754. The men are stirring the malt and water in the mash tun at the start of the brewing process. *(Credit: from The Eighteenth Century: Europe in the Age of Enlightenment, ed. Alfred Cobban, 1969, with the permission of the publisher Thames & Hudson Ltd.)*

The voyages of discovery also opened up important new supplies of fish. As noted earlier, fish have been an important source of food across the world since the earliest years of the hunter-gatherers. Rivers and lakes could only provide for a limited local population and the search for more fish was a significant driver in the development of ships and seafaring. Finding and catching the fish was only part of the problem for the fishermen, in that the catch had to be preserved if it was to reach the consumer in a fit state for the table. This problem had been solved by the Bronze Age by drying, salting or smoking, methods that served through the ages until the catch could be frozen by ice or refrigeration in modern times. By Roman times, fish had become an important item of commerce, with imports into the imperial city from Spain, Egypt and northern Europe.

Fig 2.3: Cod Fishing in North America, c.1720. The fisherman (A) with baited line for fishing (B). On landing, the cod is gutted, salted, cleaned and laid out to dry (M). Cod-liver oil is extracted in the press (I). (*Credit: from The Eighteenth Century: Europe in the Age of Enlightenment, ed. Alfred Cobban, 1969, with the permission of the publisher Thames & Hudson Ltd.*)

The observation of fast days and Lent in the Christian calendar put additional pressure on fish supply. A little could be supplied from monastic fishponds and fish traps along the coast but most fish came from large-scale harvesting of herring and cod, Yarmouth in England being noted for its herrings as early as the sixth century AD. By 1300AD the Baltic had become the most important source of salted fish, but the Dutch were soon to challenge the monopoly of the merchants of the Hanseatic League. They developed larger decked fishing vessels ('herring busses') and long drift nets that ensnared the fish by the gills. These large ships were also able to carry salters and coopers to prepare and preserve the catch at sea – forerunners of today's huge factory ships. Cabot's voyage of 1497 to the North Atlantic led to the opening up of the rich cod fishery off Newfoundland, which was to challenge that of northern Norway, which had started c.800AD in the Viking era.

The major developments in agriculture towards the end of the age of Median Technology occurred in the densely populated countries of northern Europe, where there was money to be made from further extension of the cultivated area and by achieving higher yields from crops and livestock.

One of the most significant of these initiatives was the drainage of large areas of flooded or marshy ground. The drainage works of the Romans had fallen into disuse by medieval times, and a period of wet weather resulted in the loss of much previously reclaimed land. Rising population levels in the sixteenth and seventeenth centuries prompted renewed attempts at reclamation with innovative work in Italy and the Netherlands. In Italy attention focused on the study of river flow whilst in the Netherlands methods that are still in use today were developed to reclaim lost land. Areas were surrounded by dykes built of clay, and windmill-driven scoops were then employed to transfer the unwanted water into deep canals outside the dykes.

Dutch technology was adopted in France and England, albeit in different ways. The French established a central board to direct schemes, whereas England relied on the initiatives of wealthy landowners, such as the Earl of Bedford, to bring about the extensive reclamation of fenland in Cambridge and Norfolk. This was initially achieved by cutting straight channels to discharge water that lay above sea level quickly and efficiently, with drainage windmills added later to dispose of surplus water.

Drainage apart, progress in agriculture across Europe was very uneven. Standards of living and the level of real wages in the German states had actually fallen as a result of protracted warfare in the seventeenth century, and it was only in the Netherlands and England that modest improvements in farming and diet occurred. The Netherlands was the more densely populated of the two countries at that time, and met the increased demand for food by intensive cultivation of small farms or market gardens and complex crop rotations, new sources of manure and the introduction of additional fodder crops to sustain increased numbers of animals over the winter months.

In England, market gardens were also developed around London and the larger cities and there was a modest improvement in crop yields – these still fell within medieval parameters but were sufficient to provide a surplus of corn for export by 1750. The most significant improvement was in livestock production on the lighter soils, where intensive cultivation of fodder crops increased both wool and mutton yields. Much has been made of the achievements of innovators such as Jethro Tull (the seed drill and horse-drawn hoes) and Charles 'Turnip' Townshend (fodder crops) in the eighteenth century, but their efforts have now been discredited by historians, as turnips had been introduced (from the Netherlands) in the 1630s and clover (originally from Spain) in the 1650s, but neither crop spread quickly in this period. It seems that increased use of convertible husbandry (i.e. use of temporary pastures to feed stock and manure the soil) and improvements in labour productivity were the main reasons for these advances.

A second agricultural revolution

An agricultural revolution in the UK after 1750 marks the beginning of a period of fundamental change in productivity at a time of significant increase in world population. This second agricultural revolution, if we accept that the Neolithic one as the first, enabled mankind to avoid the 'Malthusian trap' that had seemingly placed limits on population increase before this. In *An Essay on the Principle of Population* the mathematician and economist Thomas Malthus argued convincingly that population growth would always outpace sustenance, the former growing in geometrical progression (i.e. 2-4-8, etc.) whilst agricultural production could only increase in arithmetical mode (i.e. 1-2-3-4, etc.). The *Essay* was extremely influential in the years up to his death in 1834 and today his analysis is generally accepted as valid for the period prior to its publication in 1798. Subsequent progress in the second agricultural revolution was, however, to provide answers to Malthus's two major questions: 'Where is fresh land to turn up?' and 'Where is the dressing to improve that which is already in cultivation?' These developments also set in motion the massive increases in agricultural productivity that we have come to expect and sometimes worry about.

Some statistics for the century after 1750 demonstrate the impact of this revolution: wheat output in England and Wales increased by 225 per cent, the number of cattle brought to Smithfield Market in London rose by a similar amount, and there were similar increases in oat and meat production in Scotland. Such progress occurred mainly on the mixed farms on the lighter soils of the east of the UK and was due more to further advances in crop rotations and the introduction of yet more fodder crops than to any significant introduction of machinery or the application of science to husbandry. As the industrialist turned farmer Joseph Mechi put it some years later in his *How to Farm Profitably* (1864 ed.), 'the more meat you produce the more manure you make and consequently the more corn per acre you will grow in the arable portion'.

Other improvements of the period included the liming and marling of acid soils, and developments in stockbreeding that led to the introduction of new breeds of cattle and sheep. In time, advances in iron-making stimulated the rise of an agricultural engineering industry: the Rotherham plough of c.1730 was equipped with an iron-clad mould-board and the engineers Ransomes of Ipswich introduced a self-sharpening ploughshare in 1789 and an all-iron plough to replace the heavy wooden-framed examples. Iron was also used for the frames of harrows and for the tines of the wheeled cultivators that were used to break up the deep soils of land taken in during the French wars of 1793–1815. At the end of this period, however, the seed drill and horse-drawn hoe were still not in common use, and the introduction of the threshing machine, invented by Andrew Meikle in 1784, was delayed by rural unrest in the 1830s. These improvements were clearly significant, but current thinking on the causes of this British agricultural revolution tend to focus upon the wider agrarian context, particularly the replacement of a part-subsistence mentality by a concentration on the market and its opportunities for profit. This was accompanied by the emergence of the classic 'landholder/tenant–farmer/labourer' social 'pyramid' of Victorian rural society.

Progress in an age of 'High Technology'

The years after 1850 witnessed contrasting fortunes for British agriculture. Often referred to as a 'Golden Age' of 'High Farming' – 'high' meaning high standards – it was a period of significant technical change and high outputs. Many of the new machines mentioned above now came into widespread use, to be augmented by the introduction of tools to cut the hay and cereal harvests. The first mechanical reaper had appeared in 1780 but it was the horse-drawn machine with a cutter-bar and knife of the 1850s that was to bring productivity gains to harvesting. A worker with a sickle might cut a quarter of an acre in a day, whereas two men and two horses could cut, rake and bind twenty acres of wheat in the same time with the aid of a McCormick Reaper from the USA, first shown at the Great Exhibition of 1851. The productivity of the heavy clay soils was also improved by extensive schemes of tile under-drainage made possible by machine-made circular drainage tiles.

'High Farming' also involved high inputs, many bought in from abroad. Oil cake enabled farmers to keep more animals and gain more manure, whilst imports of bird manure or guano from Peru added much needed phosphates to the soil. The period also saw the introduction of chemical fertilisers such as 'superphosphates' made in alkali works where bones were treated with acid, and the gradual move from empirical methods of tillage and husbandry to those based upon science. The move from self-sustaining systems of mixed farming to those dependent upon external inputs was to become a major feature of subsequent developments.

By 1875, Britain was still 75 per cent self-sufficient in food but her farmers were now challenged by European competitors who had kept their tariff barriers after the UK had repealed its Corn Laws in 1846. Progress in France and the new nation of Germany accelerated, sugar beet was grown in increasing quantities, and Denmark specialised in dairying and pig meat for urban markets, aided by new forms of cooperative organisation. The major challenge to UK farmers, however, came from the opening up of the vast acres of the New World by railways and steamships. Some 400,000,000 acres (ten times the total acreage of England and Wales) were brought into cultivation in the US between 1860 and 1900, with further additions to the cultivated area in Canada, the Ukraine, Australia and the Argentine. These new farms were extensive and their husbandry was rather primitive by European standards but wheat could now be landed in England at a price to match the homegrown product. A shortage of labour proved a spur to mechanisation in these new lands, with the introduction of steel ploughshares and the development of a combine harvester that could harvest thirty acres of wheat in one day. More exotic foods, beverages and new ingredients such as the banana (from the West Indies in the 1870s), tea from Ceylon and palm oil and groundnuts from Africa also arrived in Europe by way of new transport networks.

These developments greatly improved the range of foods available, facilitated by advances in food preservation, storage and processing such as freezing, chilling and canning. Fishing fleets were also able to exploit new grounds in their steam-powered ships and to develop more intensive methods of trawling and drifting.

Into the twentieth century

The first part of this century was another period of transition for agriculture, with significant advances in Europe, Australasia and especially the US, where productivity rose by 77 per cent between 1910 and 1953 with 37 per cent fewer farm workers employed, whilst there was little change in other parts of the world.

A rise in standards of living in the richer countries led to greater demand for meat and poultry products and increased numbers of animals and birds. In India and China, however, oxen continued to play an important role. Improvements in breeding included adjustments to changing tastes, such as that from mutton to lamb, made easier by the introduction of artificial insemination and sperm banks. The treatment of animal disease, hitherto conducted mainly on empirical lines, became more scientifically based in the 1930s with the introduction of DDT and other pesticides to control external parasites affecting animals. Opportunities for mechanisation in animal husbandry were limited, but the milking machine was introduced in 1900 and gradually replaced hand milking. Changes in the management of animals encouraged larger farms, and more intensive methods of rearing poultry and calves were introduced, a trend which has accelerated over time.

Mechanisation became more significant in crop farming. The use of mobile steam engines was hindered by their weight but the latter years of the nineteenth century saw limited use of steam ploughing by windlass and the introduction of mobile threshing machines drawn to the farms by contractors. Small stationary engines were also employed to work barn machinery.

FOWLER'S

PATENT STEAM PLOUGHING AND CULTIVATING MACHINES,

Manufactured by CHARLES BURRELL & SONS, Licensees.

PLAN OF WORKING.—No. 1.

SELF-MOVING ENGINE, WITH PATENT CLIP DRUM.

Fig 2.4: Ploughing with Steam. The plough is pulled backwards and forwards across the land by a wire rope attached to an anchor at the opposite side of the field. Both are self-moving. (*Credit: from Charles Burrell & Sons' Illustrated Catalogue of Improved Agricultural Machinery, 1876. Author's collection.*)

Internal combustion engines were also used for this purpose, but the development of the tractor and self-propelled machines, such as the combine harvesters of the 1930s, was to bring about one of the most significant advances in world agriculture in the second half of the twentieth century.

The application of science to plant breeding produced new hybrid varieties of cereal with greater yields and resistance to certain diseases, and made crops easier to harvest with the new machines. Science also led to a greater understanding of plant

Fig 2.5: Harvesting in 1950. Five self-propelled combine harvesters at work on Crichel Down in Dorset. (*Credit: Photograph Courtesy of the Dorset County Museum.*)

nutrition and disease control and the introduction of a range of insecticides, herbicides and fungicides. The greatest breakthrough was probably the manufacture of nitrogenous fertilisers, which reduced dependence on diminishing supplies of Chilean nitrates. These advances were not problem-free – chemical fertilisers did not improve soil texture, as did farmyard manure, and some critics advocated a return to organic systems of mixed farming. Likewise, insecticides had harmful effects and the introduction of natural predators was seen as an alternative to their use.

Fishing also made use of new technology to locate and catch larger quantities of fish. Sonar and radar were employed to seek out the shoals, and larger vessels powered by steam and then diesel engines were able to travel further and to use these power sources to haul the larger nets made possible by the replacement of hemp and flax with artificial fibres. The introduction of refrigeration and large 'factory' ships enabled the fishermen to transfer their catches for processing at sea and then to return to their fishing.

A 'Green Revolution'

The 'Green Revolution', so called in the late 1960s, brought together many of these initiatives in research, development and technology transfer to achieve a massive increase in agricultural productivity across the developing world in the second half of this century. A combination of new types of hybrid cereal varieties, irrigation schemes, mechanisation, better crop management and intensive use of synthetic fertilisers and pesticides produced startling results. In 1960, India was facing famine but the introduction of semi-dwarf rice (IR8) together with these other improvements resulted in an almost threefold fall in the price of the crop and a surplus for export. Similar results were obtained in the Philippines but at the cost of a diminishing fish population in the paddy fields as the pesticides adversely affected their habitat. Such changes also had an impact in the advanced nations – yields in rice production in Japan, where farming was regarded as 'feudal' in the 1950s, have risen from 2.5 to 7 tonnes per hectare over the past century, about twice that of Bangladesh. Intensive systems of poultry and pig farming were also employed; the number of chickens kept in the world has risen from four to over thirteen billion in the past fifty years, and those killed in the US are now twice as heavy and half the age of their predecessors.

World population has tripled in the past century, with the highest rates of growth and urbanisation in the 'poor' world. That this has occurred without widespread famine and hunger is an endorsement of technical advances in agriculture, but progress has not come about without some serious reservations. Philip Lymbery's *Farmageddon: the True Cost of Cheap Meat* (2014) warns of the dangers of the wasteful, cruel and inefficient systems of mega-farming that have now come about to meet growing demands for meat. Even on the broad acres of Argentina beef cattle are now reared in factory units whilst their former grazing land has been given over to the intensive monoculture of soya to provide feed, with increasing problems for the environment such as the disposal of the increased volumes of waste from these units. The breeding of genetically modified (GM) crops poses similar dilemmas for policy-makers. Scientists have recently produced a potato that can resist the blight that led to so much misery in Ireland and other parts of northern Europe in the mid-nineteenth century. Genes borrowed from an inedible relative of this plant have the facility to recognise the onset of blight and then trigger the tubers' immune systems to resist it. This GM crop is twice the weight of an 'unmodified' one and does not require any pesticides, normally sprayed between ten and twenty-five times during a growing season. The resultant cost saving is estimated at some £72,000,000 per year. However, critics point out that there is a 'natural' species of potato that is already resistant to most forms of blight and that health risks from GM crops have been inadequately assessed. Inventors, innovators, regulators and policy-makers face many challenges as modern agriculture seeks to meet the needs of an ever-increasing demand for its produce.

Further reading

Birdsal, Derek and Cipolla, Carlo M: *The Technology of Man: a visual history*, (Wildwood House, London, 1980).

Derry, T K, and Williams, Trevor I: *A Short History of Technology*, (Oxford University Press, Oxford, 1960).

Edgerton, David: *The Shock of the Old: Technology and Global History since 1900*, (Profile Books, Exmouth, 2006).

Lymbery, Philip: *Farmageddon: the True Cost of Cheap Meat* (Bloomsbury, London, 2014).

Mechi, J. Joseph: *How to Farm Profitably* (Routledge , London,1864 ed.).

Overton, Mark: *Agricultural Revolution in England: the Transformation of the Agrarian Economy 1500–1850*, (Cambridge University Press, Cambridge, 1996).

Williams, Trevor I: *A Short History of Twentieth Century Technology: c.1900–c.1950*, (Oxford University Press, Oxford, 1982).

CHAPTER 2.1

Postscript: Milling

Owen Ward

How did man squeeze out nutrition from the natural grains which grew around him? The answer to this question is that he did so by grinding the grains into a flour, which could then be easily turned into a paste and cooked or baked into pasta or bread, such that the development of grinding skills was crucial to the history of food technology. Grinding began with teeth, and archaeological evidence of human jawbones and skulls has shown that molar teeth – aptly named – show signs of wear from such action. Then, at some time in the Old Stone Age, some ingenious persons – perhaps older men whose teeth were not what they had been – came to use a stone, with which they were already familiar as a tool or weapon, to do some preliminary crushing. This may have helped them to dust away the husks of the wild seeds that they found unpalatable, and enjoy the flour inside. Along with the best bits, however, some grit from the worn-out stones must have remained, causing the flattened molars still found in Iron Age excavations from Charterhouse-on-Mendip and elsewhere.

The work of crushing collected grain with one stone grinding on another as a saddle quern (so named because the operator stands or sits astride them) was probably women's work, while the men were out hunting or fighting. Female skeletons from the Stone Age frequently show evidence of arduous labour from crouching over a saddle quern and dragging the upper stone back and forth over a base stone. Other forms of simple crushing apparatus have included the mortar and pestle, worked by dropping a heavy log onto grain thrown into the base of a hollow tree trunk. This is a method still in use in parts of Africa, where it is a convenient form of equipment when usable stone is not available, and where water is not a reliable source of power. The origin of any ingenious new device is frequently obscure, but it is reasonable to assume that the action of rotating the upper on the lower grindstone by fixing a vertical handle to the upper stone occurred quite early to those performing the action. Also, the idea must have come to some person that it was helpful to cut a central hole in the upper stone, through which grain could be fed, while the flour so produced could be collected from the outer rim of the stones. A small gap could be created between the upper and lower stones by a spindle fixed to the centre of the lower stone and which carried a cross-piece or rynd supporting the upper (runner) stone with the appropriate gap between them. The whole assemblage required little capital, provided only that adequate stone was available.

Similar uncertainty surrounds the origins of the application of water power to the grinding operation. It is still obscure whether the simple horizontal water wheel came before or after the vertical water wheel, with the gearing that it necessitated to convert the action to a horizontal movement in the grindstones. Archaeology suggests the latter, logic the former. The advantage of providing continuous action to the grinding process without human intervention was such, however, that the application of water power in one form or the other became widespread. Left to his own devices, the miller may well have seen that the logical way to drive a horizontal millstone was to connect a central spindle to the axis of a horizontal water wheel, turning in a steady stream of water falling on one side of the wheel, and there have been thousands of them operating in all parts of the world, working to produce flour for the bakehouse and the farmyard. But the modest scale of this arrangement may well have been a disadvantage as demand increased, encouraging the construction of more sophisticated and powerful vertical arrangements, and the archaeological evidence suggests that such forms were in operation as early as, or even earlier than, horizontal mills. In addition to the more substantial construction required to house the vertical wheel and its related gearing, this form also required some modification of the stream in order to provide a good head of water and a steady supply. This involved a barrier dam to divert sufficient water into a 'leat' which delivered the water to the wheel at the most efficient height, with sluices to control the flow so that it could be turned on or off on demand.

Both vertical and horizontal arrangements of water mills were established in the Roman Empire, and from the beginning of the Christian Era the Romans took with them, along with their hungry legionaries, the skills of their milling systems as they over-ran most of western Europe. The huge millstones of Pompeii, consisting of a conical base or bed-stone carved from lava, topped by a runner-stone made in the shape of two cones point to point (rather like a sand-filled egg-timer) and turned by human or animal power, reached many of their settled urban sites. However, the most influential and productive types of Roman mill were water-powered versions of the rotary quern, using stones two or three feet in diameter, and driven by vertical waterwheels. An outstanding example of this type of mill was at Barbegal, near Arles in southern France, where the Roman engineers conducted water from the top of a steep slope, divided at the top into two parallel channels or leats, each of which led to a succession of eight vertical waterwheels some five feet in diameter, so arranged that the tail water from one mill fed into the top of the next wheel below. Such a considerable suite of mills was presumably established to feed the town of Arles, Bouches-de-Rhône, in the second century AD. The relatively flat millstones were of basalt, of an intermediary size between the 18 inch (0.46m) handmills and the later 4 foot (1.22m) or even six foot (1.83m) commercial mills, and their modest size, related to the limited size of the waterwheels that served them, could explain why Arles needed so many mills.

The original thought which enabled a vertical waterwheel to power a horizontal millstone could have developed from some industrial use, such as the need for a revolving vertical grindstone for sharpening implements. The accounts of a farrier in the nineteenth century, for an example, include almost daily entries of 'sharpening irons' – the scythes, sickles, axes and knives – which saw heavy service on the farm.

The selection of a pair of millstones appropriate for his purpose was one of the first responsibilities of the miller. His first response was probably to recover from his own fields those rocks that were hindering his own digging and sowing, and to adapt them to fit any milling apparatus he could construct. Such stones could have served for a long time, if constantly dressed with level biting surfaces. But ideally the miller would search for an appropriate quality of stone to perform the work he intended to undertake. His choice was constrained by several things. If he could identify the sort of stone he wanted, could he afford it? The ideal French burrstones for fine flour-milling cost at least twice as much as any of the alternatives. Could he get it to his mill? At least one French miller had to build a special access road for the horses to manage the slope to his mill in the valley. Was he sure that he was only going to do one kind of work? Besides grinding wheat for family bread, he may have decided also to process coarser barley or oats for farm stock consumption. So he often ended up with two pairs of stones, even if it meant having to spatchcock the second pair onto the existing machinery at a later date. The fine French burrstones would have come from the Paris basin, often sold by manufacturers in La Ferté-sous-Jouarre, Seine-et-Marne (Figure 2.1.1), which has recently taken the now defunct industry to its heart. Their products were disseminated from early in the fifteenth century as far as transport allowed, in the first instance by river, then canal, and by sea all over the world up to the twentieth century.

Other European countries also found their favoured sources of millstones. By the time of the Romans, Italy had long been quarrying the hard volcanic lava from Orvieto and elsewhere, and its use spread around Europe as the Roman Empire grew. German millers found similar sources in areas like Andernach and the lower

Fig 2.1.1: Locally-built Burrstone in la Ferté-sous-Jouarre, Seine-et-Marne, France, being examined by the Author and Nancy Raye, Maine, USA. (*Owen Ward*)

Rhine, and they were able to transport it for export, especially to Britain. England made wide use of the millstone grit found in the Derbyshire Peak District and Northumberland. Residual, unused, grindstones are still stacked near the quarries at Millstone Edge, near Hathersage in Derbyshire, but whether these particular stones were intended for milling, or for industrial grindstones, is uncertain. Ireland also found its own sources, whilst the settlers in colonial America exploited an enormous variety of rocks. The Welsh found a particularly useful source of quartz conglomerate, with the sharp grain firmly embedded in a sandstone matrix. Principally found in the Wye valley, it also occurs in many other parts of Wales. As with the French stone, distribution was possible by water, and quartz conglomerate stones are often found as the second pair, with a primary pair of French burrs. There now exists a multitude of millstone studies concerning other countries, showing how casual finds led to more invasive extraction procedures as demand grew, leaving clear evidence of careful quarrying, and even of mining.

Britain experienced a pressing need for more food for a growing urban population in the early nineteenth century, which encouraged the milling industry to shift from the countryside to new mills built at the ports that were taking increasing imports of foreign wheat and other cereals. As it was difficult to drive large establishments by water power on river estuaries where navigation and drainage were of great significance, and where wind power was too erratic to serve the purpose, steam was adopted to turn millstones. James Watt himself designed a set of beam engines to drive the Albion Mill in central London in 1784, and even earlier Matthew Wasborough had constructed an atmospheric steam engine for Edward Young's mill on Lewin's Mead in Bristol in 1780. By the late 1870s, however, these British port mills were encountering serious problems:

Many were threatened with closure. Most of these ports were now importing high grade American flour so that improving the quality of flour and cheapening production were essential to survival, while a growing demand for fine white bread acted as a trigger for promoting technological change.

One significant change to meet this challenge was the development of milling by cylindrical rolls in place of millstones. Roller mills had already been established in other industries, being widely used throughout Western Europe by the eighteenth century for milling metal sheets and bars and for crushing sugar cane. In the late nineteenth century there was much publicity about the spreading use of porcelain rolls for flour milling, which had originated in Hungary. This porcelain was not the delicate material produced for example by Chinese potteries for tableware, but was a coating of unglazed 'biscuit' porcelain with a highly abrasive surface. Such rolls eventually proved unsatisfactory and were replaced by chilled iron rolls – that is, liquid cast iron which had been poured into cold metallic moulds, 'causing sudden cooling and the formation of a very hard outer section of each roll'.

The new large mills were progressively adapted to rolls in place of millstones, but the changeover was not an easy one. When a paper by Henry Simon was read to the 1879 inaugural meeting of the National Association of British and Irish Millers wherein he advocated the roller milling system, 'his prediction of the millstone's demise met with disbelief if not total rejection from an audience of men who, without exception, utilised millstones and only those'. Perhaps the millers were unable to contemplate the sequence of processes that the roller system involved, summarised by Glyn Daniels thus: 'Rolls… opened the grain, rotating sieves separated the graduated constituents, began the process of bran removal, and sorted components into fractions graded by particle size. Air currents were used in purifiers to extract small bran particles and further rolls were employed to reduce the endosperm to flour fineness'.

The large port mills grew to include so much equipment that by the 1930s Joseph Rank's great Solent Mills in Southampton docks filled five and more floors of this huge site, more than 100 yards (91m) long and 70 yards (64m) in depth with cleaners, washers, dryers, graders, separators, and conveyors in addition to the basic purifiers, roller mills, centrifugal and other sieving devices. There were in fact two mills, one for flour and one for provender from maize. Unsurprisingly, it proved a tempting target for German air attack during the 1939–45 war, and it was partially destroyed and disabled by high explosives and incendiary bombs in November 1940. The mill was powered entirely by electricity from the town mains, suitably transformed, the first Rank mill to be so designed. The use of electricity had grown from its introduction in 1887 in Laramie, Wyoming, to rival steam power by 1920, and surpass it largely by 1940. Electric motors, and diesel engines in some cases, were employed to operate individual machines, and the new mills were then capable of processing many kinds of grain to provide varying qualities of flour and grist. Roller mills have thus come to dominate the mass market for flour, but none the less some traditional stone-milled systems survive to supply an eclectic clientele.

Further reading

Belmont, Alain: *La Pierre* à pain: *Les carrières de meules de moulins en France,* (Grenoble, 2006).

Bielenberg, Andy (ed): *Irish Flour Milling: A History 600–2000,* (Dublin, 2003).

Hockensmith, Charles D: *The Millstone Industry: A Summary of Research on Quarries and Producers in the United States, Europe and Elsewhere,* (Jefferson, 2009).

Jones, Glyn: *The Millers: A study of technological endeavour and industrial success, 1870–2001* (Lancaster, 2001).

Major, J Kenneth: 'The manufacture of millstones in the Eifel region of Germany', *Industrial Archaeological Review* (vol.6 no.3, Autumn 1982, pp.194–204).

Moritz, L A: *Grain-mills and flour in classical antiquity,* (Oxford, 1958).

Simon, Brian; *In search of a grandfather: Henry Simon of Manchester, 1835–1899,* (Pendene, 1997).

Storck, John and Teague, Walter D: *Flour for man's bread: a history of milling,* (Minneapolis, 1952).

Tucker, D Gordon: 'Millstone making at Penallt, Mon.', *Industrial Archaeology* (1971, no.3, pp.229–239 and 321–4).

Watts, Martin: *The archaeology of mills and milling* (Stroud, 2002).

Chapter 3

Power for Industry and Society

Angus Buchanan

Tools and machines

The skill of making and using tools to assist them in their activities is a feature unique to the human species, and it represents the basic form of self-empowerment by which the species has gained control over its environment. Tools for shaping and cutting, chipping and grinding, throwing and fighting, have been made of wood, bone, stone and any other available material from the beginning of human activity, and they remain indispensable for a wide range of operations in the home and workplace. Tools like the hammer and the knife are still vital for the performance of everyday functions, and are made now for many specialised uses, all of which give men

Fig 3.1: The Levant Mine. This scene at the Levant Mine, near Land's End, Cornwall, shows the proliferation of distinctive enginehouses and chimneys at what was once a busy copper and tin mining complex. (*Angus Buchanan*)

and women greater power to do and make things, thus continuing to perform as infinitely variable technologists.

Simple hand-held tools eventually developed into more complicated instruments of human ingenuity, incorporating an assembly of working parts to become machines, operating through levers, connecting rods and gear-chains to perform heavier and more laborious duties than those which could be carried out by simple tools. Machines such as mills for grinding grain, wheels for spinning the potter's clay, and lathes for shaping the carpenter's wood, were operated in the first place by the power of the human hand or foot, but were soon adapted to be driven by animal power, and then by non-animate sources of power such as the wind and falling water.

The use of such inanimate energy sources was particularly welcome for heavy and repetitive duties such as milling grain, hammering iron in the forge, felting and shrinking the fabrics in fulling mills, and stirring liquids as in paper-making and brewing. One consequence of using such power was that the scale of the operation could be substantially increased, both by individual machines and by assembling many machines powered from a single windmill or water mill. Windmills were harnessed to perform many industrial functions in flat countries such as Holland, with reliable winds and a shortage of falling water, but in general water mills were preferred because of the greater reliability of their power source. With water mills in particular it was easier to concentrate the machines being used in a single unit than it was with windmills, so that when required to do so by market conditions it was water-powered mills that tended to become large units – 'manufactories' or just 'factories' – often employing dozens, and eventually hundreds, of men, women and children. Such buildings required new thinking about the safety of large structures and promoted the development of 'fire-proof' designs using a framework of cast iron columns and beams with brick vaulting between the beams supporting flag-stone floors, the whole structure being encased in a shell of brick and glass and with slates on the roof. All these features became widely adopted in textile mills and in large buildings for civic, religious and administrative uses. In addition to increasing productivity, factories also had other advantages over individual or domestic production, because the labour force could be more adequately supervised and subjected to strict time keeping. The control of such large units thus became the genesis of modern industrial management, with work-study procedures, clock-timed control of operations, and protection for the machines, which were becoming ever larger, expensive, and vulnerable to misuse.

The Industrial Revolution

The timing of this transition to large-scale factory production was conditional on various factors in addition to their technological feasibility, which came together first in Britain in the eighteenth century. One of these was a prosperous economy, with surplus wealth available for investment in the essential sub-structure of machines and buildings, which had become available by British naval and military victories and by early successes in creating an overseas empire. Another factor was the rapid growth in financial and insurance services that accompanied the 'Glorious Revolution' of 1688, such as the creation of the Stock Exchange and the Bank of England, and the fundamental political change which took place with the triumph of a form of constitutional monarchy over the traditional notion of the 'Divine Right of Kings'. Under these conditions, a more open form of government encouraged enterprise and a degree of liberalisation.

Another important factor stimulating increased industrial activity was the vigorous spirit of enquiry generated by novel scientific speculation, beginning with the adoption of the experimental methodology advocated by Francis Bacon and by Galileo's dramatic discoveries with the telescope early in the seventeenth century. The increasing popularity of scientific enquiry was demonstrated later in the century by the formation of the Royal Society in 1662, the definition of the laws of motion and gravity by Isaac Newton, and by the confident declaration of intellectual freedom by the French philosopher and scientist, Descartes: 'I think, therefore I am.' Yet another factor was the demographic development of the period, whereby agricultural workers and their families were displaced by the enclosure programme approved by the government in order to increase the productivity of the farms, and became available as a workforce in the new factories. For most of these people it was not a happy transition, but it added powerfully to the growing industrialisation of the nation.

The outstanding factor in this transformation, which became known as the 'Industrial Revolution', however, was the emergence in the early eighteenth century of a viable steam engine. At first,

Fig 3.2: Elsecar Enginehouse. The colliery enginehouse at Elsecar, near Barnsley, houses one of the few surviving Newcomen-style steam engines that pumped water from a coal mine. It has recently been handsomely restored to its original form. (*Angus Buchanan*)

this occurred in the form of the Newcomen 'atmospheric' style, as described in Chapter 1. By the last quarter of the century, the skill of James Watt and the business acumen of his partner Matthew Boulton had converted the Newcomen design into a versatile machine capable of turning wheels and as such it was in immediate demand for many industrial applications. In particular, it quickly replaced water power in the large mill establishments springing up in many parts of the country to become the factories turning out huge quantities of textile fabrics, especially cotton and wool, and also in paper-making, ceramics, chemical production, brewing, and many other applications. It was largely thanks to the steam engine, which continued to make significant incremental improvements in performance throughout the nineteenth century, that Britain emerged as the First Industrial Nation.

It did not remain the only industrial nation for long. By 1851, when Britain was the host to the Great Exhibition held at the 'Crystal Palace' in Hyde Park, its international industrial supremacy was virtually unchallenged, but there were already signs of ingenuity and enterprise amongst the many foreign exhibits, such as American agricultural machinery, and by 1870, with America recovering from its devastating Civil War, these signs had become a formidable rivalry. Similarly, Germany by this time had become united under Prussian leadership, and by overwhelming France in the short war of 1870 had demonstrated astonishing efficiency in deploying its industrial resources for military objectives. By the end of the nineteenth century both the United States of America and Germany had overtaken Britain in iron and steel production and other indices of industrial performance, while other countries in Europe and further afield such as Japan were beginning to show that they had learnt the value of industrialisation. Britain thus surrendered its unique industrial leadership to become one of many industrial countries in a world community that had undergone an Industrial Revolution.

Throughout the period of its nineteenth century dominance in industrialisation Britain relied heavily upon steam power. Other sources of power did not disappear, and in the case of water-power it acquired a new role in the generation of electrical power by hydro-electric installations, but Britain did not possess sufficient high terrain to enable this to become an effective challenge to the steam engine. So British industry in this period depended heavily upon the use of steam engines, in both their stationary and locomotive forms. As soon as Watt's patent lapsed in 1800, engines using steam at higher pressure than Watt had considered safe became widespread, enjoying advantages of greater efficiency, as in the 'Cornish' engine designs developed by Richard Trevithick and others. The engines also became more compact, as in those for steam locomotives pioneered by Trevithick and then developed by George Stephenson and his son Robert for the first fully operational railways.

Higher steam pressures enabled designers to make their steam engines even more efficient by 'compounding' them – that is, passing the steam through two or more cylinders at diminishing pressures – and so making them more economical to run. 'Triple' compounding, with three cylinders mounted vertically and driving downwards onto the screw shaft became the ubiquitous arrangement for steam ships both large and small in the second half of the nineteenth century. The drive to ever-increasing efficiency led to many other refinements of the steam engine adding to improved performance. These included: improved valves and other working parts; 'super-heating' the steam; 'uniflow' engines with the steam passing one way through the cylinder to exit via a port in the middle, thus minimising heat loss; and 'high speed' engines, with the cylinders encased and continuous lubrication to diminish wear and tear.

The last stage of enhancement of the steam engine was the conversion from the standard linear movement of pistons in a cylinder to direct circular drive. This was achieved by the invention of the steam turbine by Charles Parsons in 1884, intended to drive dynamos for the generation of electricity, but adapted immediately for marine engines capable of driving ships at unprecedented speeds. In such uses the steam engine continued to do valuable services in the first half of the twentieth century.

By this time, however, steam power was losing its monopoly in land and sea transport, as well as in most industrial applications, to electricity and to gas or oil fuels driving internal combustion engines. Electricity requires a 'prime mover' for its generation, and in this role the steam turbine continues to give good service, although shared with hydro power and internal combustion. But the internal combustion engine has taken virtual control of road transport, in both diesel engines burning heavy oil fuels for large vehicles, and engines using lighter fuels ('petrol' or 'gasoline' or aero-fuels) in motor-cars and aeroplanes. Diesel engines have replaced steam power for locomotives on many railways, but in the long run it seems likely that electricity will become the most widely used power source for the world's railways. The efficient generation of electricity will thus become a matter of momentous concern as fossil fuels are exhausted and, despite all their dangers such as those demonstrated by the Chernobyl disaster in 1986, nuclear reactors are being considered once again as a means of providing the heat to produce the steam to turn the turbines and dynamos that generate electricity. It is to be hoped that these will safely provide sufficient electricity for an increasing consumption rate until some alternative source of power can be harnessed, such as that of nuclear fusion rather than fission.

Industrial adjustment

Meanwhile, it is certain that the industrial landscape has changed once again. The transition from a steam-power base to a base on electricity and the internal combustion engine was the first change, after that inaugurated by the steam engine. This was marked by a comparative decline of the industries which dominated the original Industrial Revolution – coal, textiles, heavy iron and steel production, engineering and shipbuilding – all of which remained important, even though they tended to shift away from Britain and Europe to more recently industrialised nations. Coal mining is determined by its geological availability and Britain is endowed with several excellent coalfields, the resources of which are by no means exhausted, even though production has fallen away from an all-time peak of 270 million tons in 1913. It has collapsed still further since the conflict between the industry and government in the 1980s led to the closure of many profitable pits, with this shift being reinforced by the increasing reliance on electricity and internal combustion as the main sources of power.

The decline of this coal industry, so recently dominant, has coincided with that of many of the industries that depended so heavily upon it, such as iron and steel, shipbuilding, and textiles, in all of which steam power had been so important. The British iron-working industry has deep roots going back at least to the Middle Ages, but it was not until the adoption of coke as the chief blast furnace fuel by Abraham Darby at Coalbrookdale in Shropshire at the beginning of the eighteenth century that its growth became vigorous. It soon blossomed in many parts of Midland Britain, especially the so-called 'Black Country' around Birmingham, together with South Yorkshire, the north east coalfield, and central Scotland. Huge factories concentrated in these areas to accommodate the large steam engines driving the rollers to turn out iron plate, rails, and bars, with batteries of blast furnaces nearby to

Fig 3.3: Coalbrookdale Furnace. The original blast-furnace adapted by Abraham Darby in 1709 for smelting iron ore with coke survived for many years in the open air. It has now been protected from the elements in a special glass tent. (*Angus Buchanan*)

smelt iron ore into iron, and with cementation furnaces and crucible furnaces to convert the iron into steel. They later acquired Bessemer furnaces to make this conversion more dramatically by passing cold air through molten iron in an open-mouth cauldron, which burnt out the impurities in a cascade of sparks, and the Siemens 'open hearth' process which performed a similar but slower transition, both introduced in the 1850s. There were also important subsidiary processes requiring frequent re-heating of the metal and hammering with James Nasmyth's steam hammer, which was itself a vertical steam engine, driving a heavy hammer-head down onto the work-piece on an anvil. Power for many of these processes was changed from steam to gas and electricity in the mid-twentieth century, but the industry contracted as it struggled to accommodate these innovations and to cope with frequent reorganisations. Even more serious was the decline in demand for the end product, as ship-building techniques changed and moved elsewhere and with

Fig 3.4: Masson Mill. One of the mill sites on the River Derwent in Derbyshire developed for cotton-spinning by Sir Richard Arkwright at the end of the eighteenth century. It has now been adapted for other uses. (*Angus Buchanan*)

many engineering industries turning to aluminium as being lighter than iron and steel.

The textile industries were hit earlier than this by a collapse in their overseas markets, after serving as one of the leading sectors of industry in Britain from the middle of the eighteenth century to the end of the First World War. It can be safely assumed, even though material evidence has long-since disappeared, that the crafts of working naturally available fibres from vegetable and animal sources were amongst the earliest industries as communities struggled to make clothes and to bind objects together. In Britain, the first sources were probably flax, a form of grass from which the fibres could be spun to make linen, and wool from sheep, which could be sheared at regular intervals. Elsewhere, cotton, from a plant, and silk, produced by silk worms, were used for the same purpose. The skills involved were first that of converting the raw material into a continuous thread (spinning) and, second, binding the thread into a fabric (weaving or knitting). Spinning was traditionally performed as a domestic task, plucking a measure of the raw material and twisting it into a skein of tightly bound fibres. This became mechanised by the use of a spinning wheel, driven by hand or pedal, which imparted the twisting motion to the skein fed into it. This long remained a task for a single spinster until, at the beginning of the Industrial Revolution, machines such as 'spinning jennies' and Samuel Crompton's 'spinning mules' were devised to work multiple spindles, and to apply water or steam power to the machines. From this point it was a simple step to combine a large number of machines taking power from a single source, and thus form the first modern factories.

Weaving was a more complicated operation and consequently more difficult to mechanise. Individual threads were initially knitted together by hand, combining two or more threads in a series of knots, but the 'loom' was soon devised as an arrangement of interlocking frames through which the threads could be passed and pressed tightly together. This frequently took the form of a hand-loom at which the operative could sit, adjusting the position of the frame carrying one set of threads (the 'warp') while feeding the 'shuttle' carrying the other thread (the 'weft') between the frames with his hands and pressing the thread into place with each passage. A further development was to provide a spring mechanism that would shoot the shuttle carrying the weft through the grid from one side to the other (the 'flying shuttle'). It took considerable ingenuity to devise a way of fully mechanising this process, but this was achieved early in the nineteenth century, and power looms were manufactured in large quantities for factory use in Britain and, increasingly, overseas.

These, together with a range of subsidiary processes all with their distinctive machines, were the main sections of all the textile industries, with a few exceptions such as the process of silk-throwing performed on rotating frames, which took the place of spinning in the silk industry. Production tended to follow a pattern of grouping in different regions: woollen cloth, for instance, flourished first in South West England (Gloucestershire, Somerset and Wiltshire) and also in Norfolk and Suffolk, but with the introduction of steam power it became highly concentrated in the West Riding of Yorkshire. Silk was less localised, but had several small regional centres such as Derby, where the Lombe brothers established one of the first factories in 1717. Linen tended to retreat to the periphery – Scotland and Ireland – with the increasing mechanisation.

Cotton, when the raw material became abundantly available from the Middle East and the southern states of the US, was adopted with enthusiasm in the late eighteenth century by Richard Arkwright and others in the river valleys of Derbyshire, but then concentrated in Lancashire around Manchester. Towns like Rochdale, Bury and Bolton acquired dozens of cotton factories, usually arranged with spinning in a multi-storey building, and weaving, because of the heavier and noisier nature of its machinery, in a single ground floor shed with north-facing sky-lights. Each of these factories had its own large chimney, indicating the presence of a powerful steam engine together with its boiler room. 'King Cotton' brought enormous prosperity to the promoters of this industry and to the country generally, even though the lot of the operatives remained harsh. It survived the famine of raw cotton that accompanied the American Civil War in the 1860s, but bounced back vigorously until the First World War. After this there was a mood of optimism that brought a lot of new investment in the industry and the installation of much new equipment, but it did not come in time to prevent a steep decline, even before the Depression of the 1930s. Thereafter the industry virtually disappeared when confronted by new problems of supply and acute competition from overseas manufacturers, except for the survival of some impressive industrial monuments.

One industry that did well out of the textile trade was engineering, which provided the essential machines and especially the steam engines for the textile mills. This industry required distinctive machines of its own, which supported many machine-tool manufacturers, all requiring lathes of various shapes and sizes, planing machines for cutting smooth surfaces, and machines for cutting, drilling, and shaping metal-work, in addition to the ever-present steam engine to activate all this machinery, and steam hammers of the type devised by James Nasmyth in the 1840s. Nasmyth's own firm manufactured steam locomotives and textile machinery before he made his fortune with his steam hammer and retired to a life as a gentleman and amateur astronomer. Firms such as this were capable of making machinery to almost any specification, which was fortunate for the engineering industry because, when the textile industry declined, new openings appeared for the manufacture of bicycles, motorcars and aeroplanes. More recently, the industry has needed to make further changes to accommodate the rapid development

Fig 3.5: Blaenavon. Several blast furnaces in various stages of decay survive at Blaenavon in South Wales. They now form part of a World Heritage Site. (*Angus Buchanan*)

of computerisation and information technology, but it has been better placed than other traditional heavy industries to take advantage of these changes.

Other industries that flourished in the British Industrial Revolution have responded with varying degrees of success to new developments in technology and market requirements. The ceramics industry, for instance, boomed mightily in the eighteenth and nineteenth centuries, with manufacturers such as Josiah Wedgwood acquiring a worldwide reputation for excellence, both in standard pottery and high-grade porcelain. The latter is a particularly fine translucent ceramic produced at a high temperature from 'China clay', or kaolin, which had been manufactured in China for several centuries before the secrets of its composition were discovered by Wedgwood and other European entrepreneurs. Wedgwood found rich sources of China clay in Cornwall that he was able to procure for his fine porcelain figurines and table-ware. This industry flourished into the twentieth century but then underwent a gradual contraction when confronted by foreign competition and is now a shadow of its former greatness, and the distinctive landscape of the 'Potteries' of Staffordshire, with its characteristic 'smog', has disappeared.

Similarly, glass making, paper-making, soap-making, and various chemical processes, have undergone a process of diminution as overseas manufacturers have undercut their markets and they have lagged behind competitors in adopting new processes. All this has involved some drastic restructuring of British industry, but it is a mistake to regard this process as a transition to a 'post-industrial society'. In fact, the need for flourishing industrial concerns remains greater than ever because the desirability of maintaining the enviably high standard of living in Britain requires that industries should continue to earn a living by producing goods that other societies are prepared to buy. Some of these will certainly be all sorts of electronic equipment that has become essential in modern households, both in the provision of machines to perform kitchen functions such as cooking, washing, refrigerating, and air-conditioning, and in providing the means of communication and entertainment performed by telephones, radio, television, and personal computers. At present, however, Britain has tended to put out many of these sorts of equipment to overseas manufacturers who can employ cheap labour to do it more profitably. Perhaps it will be necessary to reassess these priorities in order to make sure that the skills necessary to produce them are encouraged to remain closer to hand.

It sometimes appears, with the diminution of British once dominant industries, that the country is sustained by a sort of 'Indian rope trick' whereby its wealth is maintained without visible means of support. This, of course, is an illusion brought about by the collective national failure to understand fully the nature of the technological and social innovations of recent history. In reality, however, Britain has kept its place amongst the leading industrial nations by taking advantage of the scientific and technological skills acquired through high standards of education, and secured through patents and overseas contracts for designing and building all manner of structures. It has also generated a large national income from banking and other financial devices such as insurance, and even though challenged by other financial centres such as New York and Frankfurt, London continues to exercise great skill in these fields. Moreover, the high standard of its higher education system has proved attractive to a flood of overseas students who contribute substantially to British finances, and is certainly well worth encouraging as a major national service. Educational excellence has led, amongst other things, to a creditable record of valuable innovations in engineering and pharmaceutics, even though British industrialists have not always been vigorous in exploiting these commercially.

But what does Britain actually *sell*? It has long since ceased being a net exporter of coal fuels, and the exploitation of North Sea oil is a dwindling asset even though it has been enormously useful in the last four decades in keeping energy costs comparatively low. High quality steels have done much to preserve the reputation of Sheffield for maintaining the iron and steel industry, and similar high quality products in engineering – notably in the aircraft and aero-engine industries – have found good export potential. Other top range producers in the clothing and leatherwork industries continue to find profitable markets, as do a dwindling band of ceramics, glass, leather goods, and confectionery industries. It seems likely that it is in this high quality range that Britain can find the best hopes for future successes, but this implies an ever-increasing level of complex skills that can only be derived from an accompanying widening of educational competence at all levels, as well as a degree of government encouragement. The same goes for the basic industries of agriculture and fishing, which deserve support and rewards commensurate with the tremendous value of being a nation almost capable of feeding itself, even allowing for the participation in a world market on which Britain will need to depend for non-native food-stuffs. Likewise, the fish stocks of the world will need to be carefully husbanded in order to ensure continued supplies of this precious source of food. The nation learnt in two World Wars in the twentieth century that such self-reliance is a staple item in the preservation of national confidence.

New industrial developments

Another aspect of this question of national self-confidence has been a tendency to lag behind the United States and other advanced industrial countries in adopting progressive organisation and management practices. America took a lead in the 1850s in developing standardisation and mass-production in its manufacturing industries, which became known as the 'American system' of industrial organisation, even though there had been distinguished British precedents in the manufacture of replaceable parts for machines and other equipment such as Marc Brunel's block-making machinery for producing the wooden rigging 'blocks' used in abundance in the sailing ships of Lord Nelson's navy. The idea of specialisation in the manufacture of standardised goods such as pins and nails had been clearly envisaged in the first chapter of Adam Smith's *Wealth of Nations*, first published in 1775. Again, leading engineers such as Joseph Whitworth had advocated the advantages of standardised measurements and precise interchangeability of parts for machines in those that he demonstrated at the Great Exhibition in 1851. But it was then that the American application of the principle first impinged on the British public, with their small arms such as Colt revolvers and their agricultural machinery such as threshing machines, and most British industrialists seemed slow to respond to these innovations. It was a pattern which was followed subsequently with the introduction of automation in industry, adopted with enthusiasm by Henry Ford in the manufacture of his revolutionary mass-produced automobiles, rolled out from standardised parts on a moving assembly line, and priced at a range which brought them within reach of his own operatives. One consequence of the increasing automation of industry was a change in the skills of the workforce, because as processes became more dependent on machines to perform the repetitive and comparatively unskilled operations, so the emphasis shifted to workers who could control and maintain the machines. It was Ford's friend Frederick Winslow Taylor who devised procedures of 'work-study', moreover, who laid the foundations of modern managerial practice that, amongst other things, ensured the correct balance between semi-skilled and fully-skilled workers.

Following these American initiatives, other countries such as France, Germany, Japan and Britain have adopted them and adapted them to fit their own circumstances, together with an increasing automation of industrial machinery as the electronic controls of Information Technology have become available, and the industrial landscape has changed once again. The archetypical modern factory has become a light airy structure, liberated from dependence upon convenient railway stations for the delivery of fuel and raw materials, and local sources of labour by the National

Grid providing power at the drop of a switch and a network of arterial roads with motorcars available to the workforce. The workforce itself is reasonably paid, subject to agreements regularly updated, and spends much of its time supervising machinery or driving equipment. The air is clean and the tap water is drinkable. Maybe there is an over-riding uniformity of style in this, so that a factory in Swindon is very like one in Carlisle, but enlightened management can ensure that local characteristics and preferences are observed as much as possible in the detailing of the furnishings and recreational facilities. This is certainly a good working environment, and individuals will look elsewhere for variety and entertainment.

Whatever the fate of individual nations, the process of industrialisation is now firmly established in the modern world and, failing the intervention of a serious international conflict or other catastrophe, is set to continue indefinitely. The search for ever greater sources of power, and the exploration of the potentialities of new materials and processes, are being pursued with vigour all over the world, so that the demand for technological advances is accelerating as nations struggle to maintain their position in relation to their rivals. Fundamental to the question of national self-reliance and independence is the problem of energy supply. If Britain is to remain 'in business' as a leading industrial nation, it has to solve the problem of acquiring sources of power which are both safe and reliable, and the exhaustion of its sources of fossil fuels cannot be delayed for more than a few more decades. Engineers are already finding both traditional and ingenious new ways of harnessing wind and water power; traditional, in such devices as wind turbines, direct descendants from the ancient windmill; and novel, such as in tidal barrages and wave-generation, and tapping the heat below us in the Earth's mantle. The disruption and enormous capital cost of some of these projects are deterrents to rapid decisions on such enterprises, but the increasing rarity and cost of fossil fuels may eventually make expensive alternatives more realistic propositions, and the most accessible of these is atomic power.

Atomic Energy

The Second World War ended abruptly in August 1945, when atomic bombs were detonated over the Japanese cities of Hiroshima and Nagasaki. Until then, it had seemed likely that the war, which had already gone on for six years, would be indefinitely prolonged by the determination of Japan to resist the humiliation of defeat at the hands of the Western Allies and Soviet Russia, who had so recently defeated Nazi Germany. However, the horror of the devastation caused by these new weapons was so great that the Japanese hastened to surrender in order to avoid the infliction of similar damage throughout their heartland. The victorious Allies were subsequently so appalled by the scale of the destruction caused by their release of atomic power as a weapon, and its long-term effects in terms of radiation pollution and a lengthening casualty list, that some of them came to have serious reservations about the wisdom of their action. So far, these two bomb blasts have remained the only instances of the use of the new weapons against human targets.

The atomic bomb was one of the results of the discovery of radiation from certain heavy metals at the end of the nineteenth century, and its immediate impact on medical practice through the use of X-Rays. It also inspired Rutherford's 'planetary' model of atomic structure and the theoretical possibility of splitting the atoms of certain materials, unleashing a tremendous explosive energy. It took the crisis of the war in 1939, however, to promote a race to harness such energy in a usable bomb, achieved by the Western Allies in the summer of 1945 and immediately put into effect. Ever since, the nations of the world have lived under the shadow of a general atomic war, which has served to keep the most serious conflicts within the limits of a 'Cold War'; and gradually the states possessing this power have come to accept a partial code for its control.

Meanwhile, the peaceful use of atomic power to generate energy through an 'atomic pile' in which the nuclear reaction is controlled to release heat gradually, raising steam to produce electric power through turbines, has been widely explored and put cautiously into practice, taking some of the pressure off the world consumption of fossil fuels. There have been many difficulties, in the preparation of atomic fuel, in using it safely, and in disposing of the radioactive waste. The most serious accidents so far have been at Chernobyl in Ukraine in 1986, due to human error, and at Fukushima in Japan in 2011, when the sea defences were overwhelmed by a tsunami induced by an earthquake, both coming close to a disastrous leak of radiation. The development of the peaceful application of atomic power has thus been very uneven and continues to cause serious concerns about its safety, and it seems likely that these anxieties will only be overcome by the achievement of control over nuclear fusion rather than fission. This is theoretically possible but it requires maintaining extremely high temperatures for more than a few seconds, and ways of doing this have still to be discovered. The rewards will be high, as fusion of simple elements in sea water should provide abundant cheap fuel while avoiding harmful waste products.

In addition to the still problematic power sources in atomic fission and fusion, the Second World War encouraged other significant technological innovations, both by accelerating discoveries made earlier, or by new discoveries. In the field of artificial materials such as 'plastics', derived from the molecular manipulation of carbon-based materials in coal and oil (as distinct from traditional 'plastics' such as rubber), important uses were found for 'celluloid' such as in photographic film, for 'bakelite' in solid objects such as wireless cases, and for 'nylon' in parachute fabrics and ladies' stockings. Other new materials have been introduced since the war, including metal alloys for a diverse range of applications, culminating in carbon fibre steel, giving step-change advantages in the weight / strength ratio of enormous importance in aviation engineering

In pharmaceutics, antibiotics such as penicillin, neglected since its discovery by Alexander Fleming in 1928, were developed dramatically as a medical aid during the war, as were other drugs and insecticides. The war also stimulated a wealth of life-saving enhancements in medical equipment, leading to the modern machines for scanning the internal organs of patients and for enabling extremely complex operations for replacing such organs.

In aeroplane engineering, the jet engine, invented by Frank Whittle in the 1920s and developed slowly by him and other European inventors in the 1930s, only came into successful operation during the war, as did the helicopter. Whereas the internal combustion engine powered by petrol or diesel fuels has retained its dominance of road traffic throughout the twentieth century and appears likely to maintain such dominance well into the present century, railway traffic has undergone substantial development. Steam traffic survived the war, but was then largely replaced by electric traction. Diesel engines have enjoyed a relatively brief lead on many railways, but there seems to be little doubt that most world main line railways will continue the well-established trend to electric traction. 'MagLev' – the use of powerful magnets to cause the levitation of trains so that they float above a prepared track and then drive the train along it – has been under slow development for several decades, but has recently enjoyed some operational success in Japan, Germany, and elsewhere. But its high costs of installation compared with traditional track have deterred many operators, despite its acknowledged advantages of maintaining high speeds and safety. It would seem to have promising future prospects.

Since the War

The D-Day invasion of continental Europe by the Western Allies in 1944 was greatly facilitated by ingenious ideas for mobile harbours and pipelines. And even though on the point of defeat, the Nazis with their 'vengeance' weapons, the V-1 and the V-2, inflicted serious damage on Britain. The V-1 was an effective flying bomb which followed a pre-directed course until it ran out of fuel, while the V-2 was a missile rocket from which the rockets of the space age have all been derived. These rockets constitute a new sort of prime mover, deriving their propulsive energy from the reaction of burning volatile fuels, usually in liquid form, in a controlled manner. They have become indispensable weapons of modern warfare, although in their largest form – as inter-continental ballistic missiles – they have not yet been used operationally. Most dramatically, however, they have performed impressively in the exploration of space beyond the Earth's atmosphere, including landings on the Moon and journeys to other planets in the solar system.

Yet another successful technological development stimulated by the Second World War, but by no means the least significant, the invention of 'radar' by British engineers as a radio-detection device immediately before the war has had momentous consequences. The world electronics industry was then in its infancy, but through the emergence of radar and radio stemmed the modern 'computer', which was devised initially for code-breaking machines but which has transformed modern life and industry. The computer itself is traditionally derived from the remarkable nineteenth century partnership between Charles Babbage, creator of the 'Difference Engine' as a calculating machine and the 'Analytical Engine' for solving equations, and Ada Lovelace, a mathematical genius who devised a digital formula enabling such machines to be programmed.

In practice, these machines were too complex to be built with the technology available in the mid-nineteenth century, and it was not until a century later that the development of electronic valves made it possible. Then Alan Turing, working at cracking enemy codes at the secret government establishment in Bletchley

Park, devised *Colossus* in 1943, the first effective realisation of the Babbage–Lovelace aspiration. Soon after the war, the invention of the 'transistor', a basically simple and robust combination of semi-conducting metals capable of detecting a wireless signal, replaced thermionic valves and made possible a gigantic step towards the miniaturisation of the computer, printed circuits, micro-miniaturisation, and the near miraculous capacity revealed by 'nanotechnology' – the new science of manipulating matter on an atomic or molecular scale. One example is the extraordinary capacity of the 'memory stick', several of which can be comfortably held in the human hand. Computers shrank from gigantic machines like *Colossus* to hand-held iPhones and tablets. It has been a spectacular revolution.

All this equipment, and much more, such as lasers ('Light Amplification by the Stimulated Emission of Radiation') – invented by Theodore Maiman in 1960 – and a huge number of other appliances, created an ever-increasing demand for power to drive them. Perhaps in the long run the solution to the power requirements of world society will depend upon some direct access to solar power without the space-hungry use of panels of solar cells. Somehow or other, in order to survive, the sophisticated industrial society of the future will need to have found a successful solution to this problem. The speed of technological progress in recent decades makes the prospect of a realisation of this goal in the present century a reasonable expectation, and will fully justify our use of the concept 'Technological Revolution' to describe the transformation of the modern world since the beginning of the eighteenth century.

Further reading

Brown, J A C: *The Social Psychology of Industry* (Penguin, London, 1954).

Buchanan, R A: *The Engineers: A History of the Engineering Profession*, (London, 1989).

Landes, David: *The Unbound Prometheus* (Cambridge, 1969).

Mathias, Peter: *The First Industrial Nation*, (Methuen, London, 1969).

Pollard, Stanley: *The Genesis of Modern Management*, (Penguin, London, 1968).

Rolt, L T C: *Tools for the Job: A short history of machine tools*, (Batsford, London, 1965).

Chapter 4

Structures – Buildings and Civil Engineering

Stephen K Jones

From the beginning of time human society has endeavoured to change and adapt the environment around it. Such constructional change was initially based on empirical evidence, but from the Renaissance the physical science underlying structural engineering began to be understood. An understanding of how such structural material behaved was largely unknown prior to the Industrial Age. In the sixteenth century Leonardo da Vinci produced engineering designs despite the lack of beam theory and calculus, based on scientific observations and discipline, and in the following century Galileo Galilei, Robert Hooke, Isaac Newton, and Gottfried Leibnitz made contributions to the scientific foundation of modern engineering. In the eighteenth century the work of Leonhard Euler gave engineers the methods to model and analyse structures and, in conjunction with Daniel Bernoulli, he devised the Euler–Bernoulli beam equation – a major advance in structural engineering design. In 1757 Euler derived his buckling formula, which would greatly advance the ability of engineers to design compression elements. Materials science and structural analysis would play a major role in industrial change by enabling new structural materials to be brought forward during the Industrial Revolution. In this period, development in iron and steel transformed Britain from a mainly agrarian economy to an industrial one with an understanding of the fundamental nature of materials.

There was also an important step forward in the practice of civil engineering, with the emergence of a professional body: the Institution of Civil Engineers (ICE). Established in 1818, the ICE represented a 'new' profession and can claim to be the first of the industrial age, indeed, the first profession since the ancient ones of law, medicine and clergy. The Industrial Revolution also saw the parting of the ways of the engineer and the architect into separate professions and the rise of the professional engineer during the nineteenth and early twentieth centuries. The development of structural theories and the specialised knowledge and application of engineering materials changed the face of the built environment forever. It was an environment that was constantly changing with the greater availability of materials, for example: refractory metals, light alloys, plastics, and synthetic fibres in the twentieth century. These new materials were used because they are better, more versatile or cheaper than those available during the nineteenth century, or have combinations of properties that can realise new designs in engineering. Another great advance has taken place with the advent of digital computers from the 1950s, enabling the development of finite element analysis as a significant tool for structural analysis and design. Such advanced analytical methods and the rise of computer modelling continues the great advance in structural engineering design.

Traditional materials: stone, brick and timber

Stone has a long history as a widely used building material, from ancient times through to the present. Most buildings reflected the local character of stone, with many small towns and villages having their own local quarries, the diversity of colour and texture reflecting local geology. Prior to the Industrial Age stone was not hauled great distances except for major projects such as large ecclesiastical buildings and castles where Norman builders retained their preference for limestone such as Caen Stone. Brick (and mortar) has been an effective building material for a similar period but like stone is lacking in tensile strength, resulting in distinctive arch designs. Brick is an artificial stone formed of clay moulded in rectangular shapes of constant dimensions and hardened by burning or exposure to the sun. It was not until 1776 that standard sizes were prescribed by the Bricks and Tiles Act, but excise duty on these was applied until 1833. By the mid-nineteenth century the impact of the railways, both in the need for brick structures and in facilitating the transportation of bricks, led to a massive increase in brick making and by the end of that century machine-made wire-cut bricks were being used in great quantities.

Timber also has a long history as a building material and at the start of the Industrial Revolution was in considerable demand for bridge building. Despite the predominance of stone, the increasing use of brick arches, and the introduction of 'new' materials such as cast iron, wrought iron and steel, wood continued to be used throughout the nineteenth century with a great number of railway viaducts and bridges being constructed in timber. However by the eighteenth century native supplies of domestic hardwood, such as oak, were becoming scarce, and following the trend started by shipbuilding, softwood timber from the Baltic states began to be sourced. Baltic pine was far cheaper than iron or steel and it could also be worked on site without the need of a forge. Brunel would make extensive use of timber for his viaducts and bridges in the 1840s, designing the timber elements to be replaceable in service – a degree of structural 'over-engineering' that allowed the replacement of decayed or defective parts. His timber truss structures, particularly the 'fan' types, used propped multi-span beams to create the trestles of the viaducts. Other engineers such as William Cubitt, Thomas Longridge Gooch, John Green, Robert Stephenson and John Sutherland Valentine also designed timber viaducts, some of which employed laminated arches.

Cast iron

At the end of the eighteenth century, both cast and wrought iron were becoming more readily available and playing an increasing role in structural engineering. They have distinctly different properties, with cast iron being strong in compression but weak in tension. Wrought iron is highly ductile and strong in both tension and compression. In 1769 a 72ft (21.95m) span wrought iron bridge at Kirklees in West Yorkshire preceded the construction of the famous cast iron arch at Ironbridge in Shropshire, built in 1779 by Abraham Darby III, the iron founder of Coalbrookdale. Ironbridge is the oldest surviving cast iron bridge, and this form became very popular in the next sixty years, although it would prove to be unsuitable for large span railway bridges. Cast iron was also used in the construction of buildings. The next large cast iron bridge, Wearmouth Bridge, opened in 1796, the design of which is sometimes attributed to the radical writer Thomas Paine but the design was actually finalised by Rowland Burdon with advice from architects and engineers. It had a span twice as long as Ironbridge at 240 feet (73.15m) and at the time of building was the biggest single span bridge in the world, and high enough to allow ships to pass beneath without lowering their masts. The bridge underwent major repair in 1805 and in 1859 was reconstructed by Robert Stephenson.

Fig 4.1: The first Wearmouth Bridge photographed c.1854 by Edward Backhouse (1808–1879). The current bridge is the third Wearmouth Bridge in this position, a through arch bridge that opened in 1929. (*Photograph courtesy of and copyright of Sunderland Museum & Winter Gardens.*)

William Strutt built a textile mill at Belper in Derbyshire in 1792 using cast iron columns carrying timber floor beams protected against fire by plaster. Ditherington Flax Mill in Shrewsbury, built in 1797 to a design by Charles Bage, was the first building in the world with an interior iron frame. Bage called upon the iron founder William Hazeldine, whose Shrewsbury works cast the columns and beams. Strutt and Bage later collaborated in the construction of Belper North Mill with a full cast iron frame – the first fully 'fire-proofed' building in the world. A few years before this saw the construction of the first cast iron tramroad bridge, Pont-y-Cafnau, across the River Taff at Merthyr Tydfil. With a 48ft (14.63m) span it was authorised in January 1793 and almost certainly built by the Cyfarthfa ironworks engineer, Watkin George, it still stands. There would be a tendency for engineers to overdesign due to the lack of tensile strength in cast iron, for which wrought iron was found to be superior.

Wrought iron

Wrought iron is literarily 'worked iron', and as such the traditional material of the blacksmith, who would work the iron repeatedly under a hammer. The earliest form of wrought iron was 'charcoal iron', which as the name suggests was made in a charcoal fire and was used from the Iron Age to the end of the eighteenth century. 'Puddled Iron' was made from the reworking of cast iron in an indirect coal fired furnace and was used at the start of the industrial revolution. Wrought iron is softer to forge and is more malleable and weather resistant than its modern equivalent, mild steel. Wrought iron would be taken up in earnest for larger spans, particularly with the development of iron chain suspension bridges at the turn of the nineteenth century. The first major bridge of this type was the Union Chain Bridge by Captain (later Sir) Samuel Brown RN opened in 1820. Spanning the Tweed between Northumberland and Berwickshire, it was the longest wrought iron suspension bridge in the world at the time, and the first vehicular bridge of its kind in Great Britain. The bridge was built by Brown who patented his design for the wrought iron chains that were made at Brown's Pontypridd chainworks. At 437 feet (133.2m) in length between the suspension points and 18 feet (5.5m) wide it is the oldest suspension bridge in the world still carrying vehicular traffic. Brown had previously worked with Thomas Telford on an earlier, uncompleted, design for bridging the Mersey at Runcorn

Fig 4.2: Union Chain Bridge, a Grade I listed building and a Scheduled Ancient Monument. (*Photograph by Stephen K Jones*)

and his chain design influenced Telford's subsequent suspension bridge design at Menai and Conwy.

The flexibility of the suspension bridge, however, made it unsuitable for railway use and increasing demands for wider spans created a demand for wrought iron girder and truss bridges. The first wrought iron bridge was built for the Pollock and Govan Railway near Glasgow by Andrew Thompson in 1832. Radical development however led from the mid nineteenth century work by William Fairbairn to the tubular railway bridge by Robert Stephenson. Brunel also developed his tubular suspension bridge at the same time and although seen as a forerunner of his Royal Albert Bridge, the Chepstow Bridge was essentially a prototype Pratt-type truss design. One of the last large structures to be built of wrought iron was the Eiffel Tower, constructed by Gustave Eiffel and Maurice Koechlin in 1889.

Steel

Both forms of iron would be overtaken from the mid nineteenth century by steel. Steel construction for large structures became practicable in the 1850s when Henry Bessemer invented a process

Fig 4.3: Forth Railway Bridge, the first major steel-based construction project in Great Britain. The steel was supplied by the Siemens Landore Works and the Steel Company of Scotland. Photograph by George Washington Wilson (1823–1893). (*Courtesy of the Institution of Civil Engineers*)

for manufacturing steel by blowing cold air through molten iron in a 'converter'. At about the same time, William Siemens perfected his 'open-hearth' process, and together these two developments made mild steel available in bulk. Despite some early brittle fracture failures in structures fabricated from mild and low alloy steel it gradually replaced both cast iron and wrought iron in structural work. The first major railway bridge to be built in steel was the Kymijoki railway bridge over the river Kymi in Finland in 1870. In the US the Eads Bridge in St Louis, designed by James Buchanan Eads and opened in 1874, was the first alloy steel bridge. The greatest statement of the steel age was the Forth Railway Bridge, with its three huge cantilevers. It was engineered by Benjamin Baker and Sir John Fowler with the contractor William Arrol and opened in 1890. At the time it was the longest single cantilever bridge span in the world with two separate spans of 1,710 feet (521m) and today the second longest behind the Quebec Bridge in Canada at 1,800 feet (549m). After eight years of construction the 1.5 miles (2,467m) long bridge over the Firth of Forth was completed in 1890.

Concrete

Portland cement was patented in 1824 by Joseph Aspdin as 'a superior cement resembling Portland Stone'. Forms of cement had long been in existence and 'Roman' cement had been in common usage in Europe from the 1750s. Aspdin's patent enabled cement to be made cheaply from commonly available materials and increased the range of building applications. In 1848 Joseph-Louis Lambot patented his system of mesh-reinforcement and concrete to make a boat out of ferrocement in 1855 – the forerunner of modern reinforced concrete. Joseph Monier took the idea forward using steel-mesh reinforcement and filed several patents for tubs, slabs, and beams, leading eventually to the Monier system of reinforced structures, the first use of steel reinforcement bars located in areas of tension in the structure. From 1892 onwards François Hennebique's firm used his patented reinforced concrete system to build thousands of structures throughout Europe. The Weaver & Co Flour Mill in Swansea built in 1897 was the first British, and Europe's first, multi-storeyed fully framed reinforced concrete building. The main patent for the Hennebique system dates from 1897, the year the mill was completed, under the eye of Hennebique's British agent, Louis Mouchel. Mouchel promoted the system for a wide range of structures, including jetties, bridges

Fig 4.4: Weaver's Mill, designed by the French engineer Français Hennebique and demolished in 1984. Photograph taken 1979. *(Copyright Brian Whittle and licensed for reuse under this Creative Commons Licence)*

and retaining walls. Some 130 concrete frame Hennebique-system buildings were erected in Britain by 1908, considerably more than by any other system. The first reinforced concrete bridge erected in Scotland was a 28ft 8.53m) span road and rail bridge in Dundee (1903).

In 1899 Wilhelm Ritter formulated the truss theory for the shear design of reinforced concrete beams and Emil Mörsch improved this in 1902. Reinforced concrete came into its own as a structural form through the pioneering work of Robert Maillart, whose attention to detail resulted in the revolutionary Salginatobel Bridge across an alpine valley in Schiers, Switzerland, opened in 1930. Prestressed concrete, applying tension to overcome the tensile weakness of concrete structures, was pioneered by Eugène Freyssinet and patented in 1928. Freyssinet had constructed an experimental prestressed arch in 1908, and later used the technology in a limited form in the Plougastel Bridge in France in 1930. Today concrete is the most widely used man made material.

Newer materials: glass, aluminium and plastics

By the seventeenth century blown plate glass was manufactured by a very labour-intensive process grinding broadsheet glass. This gave way to the manufacture of polished plate glass in the next century, and later the introduction of steam powered machine grinding and polishing enabled large panes of very good quality glass to be made. In 1834 a German improvement of the cylinder sheet process led to the production of much larger sheets and better quality glass and the main means of manufacturing window glass well into the twentieth century. The end of duty on glass in 1845 saw a great increase in demand and the price dropping by up to 75 per cent. The potential of glass for new structural designs was developed by British manufacturers such as Chance Brothers, one of the earliest to employ the cylinder process in Europe. This was followed by a radically new type of building by Joseph Paxton: the 'Crystal Palace'. This plate-glass and cast-iron structure, supporting timber floors, provided an exhibition space of 990,000 square feet (92,000m^2). However, as a design style it would be largely limited to exhibition halls and for railway station roofs.

The Home Insurance Building in Chicago was completed in 1884, the first tall building to use structural steel as a framework. At 180 feet (55m) high it was very light compared to contemporary brick buildings and is usually considered the first real skyscraper. Originally built as the Fuller Building, the Flatiron, a triangular twenty-one-story 307 feet (94m) high steel-framed building, is

Fig 4.5: The Crystal Palace, exterior view of the west end of the building, photographed 1851, from *Reports by the Juries*, (1852), Spicer Brothers, London. Photograph by Claude-Marie Ferrier or Hugh Owen. (*Courtesy of US Library of Congress Prints and Photographs Division, Washington, D.C. LC-USZ62-63009*)

Fig 4.6: The Flatiron Building, New York, under construction in 1902. (*Courtesy of US Library of Congress Prints and Photographs Division, Washington, D.C. LC-D401-14278*)

located at 175 Fifth Avenue, Manhattan, New York City. Designed by Daniel Burnham in the Beaux-Arts style it was built by the George A Fuller Construction Company for their headquarters. One of the tallest buildings in New York, it is a 'skyscraper' of innovative steel skeletal design that carries the entire load of the walls. A limestone and terra-cotta façade incorporates classical Roman features into a modern building with sculpted decoration. Another innovative feature was the six hydraulic elevators installed by Otis. In 1903 laminated glass was invented by accident and would be developed with a thin plastic film between two sheets of glass; this allowed much larger windows with improved safety and security, much of which could be glazed undivided by glazing bars. This would lead to glass 'curtain walling' designs for high-rise buildings and today glass strongly influences modern architectural design. Steel frames would be an important element of modern high-rise constructions in the twentieth century and were employed by architects such as Fazlur Khan. The DeWitt-Chestnut Apartment Building in Chicago was the first building to apply the tube-frame construction designed by Khan, and this laid the foundations for the tube structures used in later skyscrapers, including the World Trade Centre.

After steel, aluminium is the second most widely specified metal for building and is used for window frames, roofing, cladding and curtain walling as well as for prefabricated buildings. Aluminium is a 'new' metal, first extracted and refined from the principal ore, bauxite, with commercial output beginning in 1854. It offers an exceptional strength to weight ratio, being 66 per cent lighter than steel and less susceptible to brittle fractures. The cladding of the dome of Rome's San Gioacchino's Church in 1898 is widely cited as its first structural use and it would be used to great advantage in the Art Deco inspired skyscrapers of New York. Aluminium's great advantage for civil engineering was taken up in Pittsburgh with the first aluminium bridge deck in 1933, a replacement of the steel decking of the Smithfield Suspension Bridge. The first all-aluminium bridge was the 100 feet (30.5m) wide riveted aluminium plate-girder Grasse River Bridge for railway traffic in 1946 at Massena, New York. In Europe the first aluminium bridge dates from 1949. In 1955 the De Havilland hangar at Hatfield used aluminium alloy trusses to give a clear span of 200 feet (61m).

Whilst composite materials can technically include bricks, made with clay and straw, and cement, the first modern composite material was fibreglass. As well as boat hulls and car panels its use as building panels forms part of modern architectural design with plastics playing a major role for the decoration of the building interior and pipework.

Engineering: canals, harbours, docks and lighthouses

As an island Britain relied on its ports and harbours to facilitate external trade and commerce. Important elements were the opening up of inland navigation through the improvement of rivers and the cutting of canals and the construction of docks and lighthouses. From a Roman heritage of navigable cuts known as 'dykes' to bypass navigational hazards in rivers and to improve drainage, Britain's first canal with locks was the Exeter Ship Canal commenced in 1564 and opened in autumn 1566, by John Trew of Glamorgan. Leonardo da Vinci's design of mitre lock gate, of 'V' shaped gates forced together by water pressure, was also used on the River Lee, at Waltham Abbey in 1574. The first canal in Britain to be built without following an existing watercourse was the famous Bridgewater Canal opened in 1761 and engineered by James Brindley. With the canal boom that followed, the demand for engineers drew in men from many other disciplines, such as mining engineers and millwrights who were used to working with water for waterpower. Brindley went on to engineer the Trent and Mersey Canal and would be responsible for over 300 canal projects. He set the standards for most of the canals that followed, especially the narrow canals, and his many assistants would go on to become well known canal engineers. Later engineers such as Thomas Telford would move away from these early 'contour canals' to engineer more direct canals using cuttings and embankments.

Ports and harbours would provide much work for the engineer in the nineteenth century; indeed they proved almost a continuous workload for the engineer during the nineteenth century due to the increasing sizes of ships. Floating harbours, which had developed at the river entrance to the sea, were giving way to wet docks with single or double lock entrances. Greater access for loading and unloading, without the vessels having to 'take the ground' at low tide, enabled a greater turnaround of shipping. Extensive enclosed wet dock systems grew up at ports such as Cardiff, Liverpool and London. Britain's world trade was a strong imperative to push for the expansion, improvement or even replacement by new construction of existing docks and harbours. It was not just the ever-increasing size of ships that forced this pace, but the increasing dominance of the steamship and its high capital costs demanded ever greater turnaround at the dockside to ensure maximum utilisation. Docks would also be developed in locations previously considered unsuitable for access by sailing ships but which could now be facilitated by steam tugs in unfavourable winds.

The first person to describe himself as a civil engineer was John Smeaton in 1761. Smeaton was responsible for the reconstruction of the Eddystone Lighthouse and was involved in canal and river surveying, undertaking his most successful work in Scotland on the Forth and Clyde Canal and in Ireland on the Grand Canal with William Jessop. The son of a naval shipwright, Jessop's father Josias had worked with Smeaton, who also took on William for training as an engineer. Jessop worked on river navigations and canals and he was engineer on the Grand Junction (Grand Union Canal), the Ellesmere Canal, the Rochdale Canal, the East India docks in London and dock improvements in Bristol. On the Ellesmere Canal there would be an uneasy relationship with Thomas Telford. Telford began work as a stonemason and undertook various architectural works before being appointed Surveyor of Public Works for Shropshire. He moved into canal works and on the Ellesmere Canal, under Jessop, he was responsible for the iron aqueducts at Pontcycyllte and Chirk. Further canal work followed in Britain and Europe, with much harbour construction in Scotland, whilst his road building work, such as on the London to Holyhead road, would result in the Menai Suspension Bridge and the Conway Suspension Bridge on the Chester road. Both suspension bridges drew upon William Hazeldine's iron-making expertise in terms of the wrought iron chains. John Rennie, who also began work as a millwright after setting up as a London-based engineer, became the engineer for the Kennet and Avon Canal and worked on many dock and harbour schemes.

Engineering: mining, land drainage and tunnels

As well as the building of roads, railways, canals and bridges, engineers were also engaged in mining, the driving of tunnels and land drainage. Before the eighteenth century much civil engineering expertise in Britain and Ireland was derived from elsewhere in Europe, such as German mining practices. Few British engineers worked abroad, but one early major British overseas work was the

Tangier Mole of 1667–1680. In 1782 a Soldier Artificer Company was established in the British Army for service in Gibraltar and was responsible for the great network of tunnels and defensive works on the 'Rock'. Such skills were similar to driving horizontal or near-horizontal adits or passages into the side of a hill or mountain for the purpose of working, ventilating, or removing water from a mine. To win minerals from deeper levels vertical shafts were driven, demanding ever increasing depths in which the shaft would be driven to extract coal in the nineteenth century. Land drainage works required extensive capital and were often undertaken by Adventurers, who in exchange for parliamentary sanction, gained rights to the land reclaimed as a result of the civil engineering works. Major schemes included the draining of the Bedford Levels in the Fens district. The Dutch engineer Cornelius Vermuyden was employed to carry out the work, and he constructed an elaborate system of drains to drain the marshes. Whilst such schemes were not without criticism by those with common land rights, between 1760 and 1840 most of the fens were drained and enclosed by Act of Parliament. However, as the land dried out it shrunk, becoming lower against the water table and more susceptible to flooding. Pumping had to be introduced, at first by windmills, then by steam engines, but the result brought forward more arable land.

Pre-history examples of tunnels can be found around the world and, of course, tunnelling had been practiced for many centuries by miners, but more strategic tunnels such as sub-aqueous tunnels would be put forward in the late eighteenth century. These included Ralph Dodd's proposal for a tunnel under the River Thames, following on his earlier proposal for a tunnel under the River Tyne. Another project was put forward for the Thames by the Cornish engineer, Robert Vazie, who proposed a tunnel from Rotherhithe to Limehouse. This led to the formation of the Thames Archway Co., which was incorporated by an Act of Parliament in 1805. Construction of this tunnel was fraught with difficulties and floodings, and eventually, another Cornish engineer, Richard Trevithick, was called in by Vazie but this attempt was to end in failure. Marc Isambard Brunel now attempted to drive a tunnel under the Thames and set out to accomplish this task through an innovative approach, the use of a tunnelling shield to protect the miners working at the face. Construction started in 1825 but the river did break in on a number of occasions. The most disastrous influx occurred in 1828, when six men drowned and the young

Isambard Kingdom Brunel, acting as resident engineer, just escaped with his life. The tunnel was pumped dry; but the cost of making good this, the second major flooding, exhausted the funds of the Thames Tunnel Co. After an interval of seven years work resumed and, in 1843, the tunnel was finally completed but without access for horse-drawn traffic. Today the tunnel forms part of the London Underground system.

Further reading

Timoshenko, Stephen P: *History of Strength of Materials: With a Brief Account of the History of Theory of Elasticity and Theory of Structures,* (New York: 1983).

Billington, David P: *The Tower and the Bridge: The New Art of Structural Engineering,* (Princeton: 1985).

Blockley, David: *Bridges: The Science and Art of the World's Most Inspiring Structures,* (Oxford; 2010).

Collins, Peter: *Concrete: The Vision of a New Architecture,* (Montreal: 2004).

Cossons, Neil, and Trinder, Barrie: *The Iron Bridge: Symbol of the Industrial Revolution,* (Chichester: 2002).

Paxton, Roland: *Dynasty of Engineers: The Stevensons and the Bell Rock,* (Edinburgh: 2011).

Rolt, L T C: *Isambard Kingdom Brunel* (London: 1957) and *Thomas Telford,* (London: 1958).

Skempton, A W: *Civil Engineers and Engineering in Britain, 1600–1830,* (Aldershot: 1996).

CHAPTER 5

Transport I – A Brief History of Shipping

Giles Richardson

More than 70 per cent of the earth's surface is covered by oceans, lakes and rivers. From the earliest prehistory these bodies of water have acted as both obstacles and highways as mankind has attempted to gain mastery over them. The development of watercraft enabled humans to explore and exploit the environment, to migrate and colonise new areas and to facilitate trade, travel and war between peoples and civilisations. The study of the history of technology would therefore be incomplete without some understanding of the evolution of shipping and mankind's interaction with the maritime world. This chapter will focus on the archaeological remains of ships and boats to outline the development of maritime technologies from the earliest seafaring until the start of the twentieth century. With such a wide historical and geographic range it would be impossible to cover all regions and periods in equal depth, and this chapter will share the bias of archaeological evidence that is strongly based on the Mediterranean and Northern European regions.

Prehistoric voyages

When the earliest use of seaborne transport took place is unknown, but almost all major barriers of water were crossed at a remarkably early date. The Indonesian Island of Flores has revealed evidence for occupation by the hominid species Homo floresiensis (Flores Man) from about 100000BC. These first inhabitants must have arrived by sea, crossing 12 miles (19.3km) from the mainland during the last glacial period. In Europe, trading voyages were occurring by 11000BC as sailors in the Mediterranean shipped the volcanic glass obsidian between its source on the island of Melos and mainland Greece, where it is found in settlement sites worked into tools such as blades and scrapers. Bones from deep-water species show that

Mesolithic populations were carrying out offshore fishing in the Mediterranean Sea by 6000BC.

The very first watercraft are likely to have been simple types of raft, constructed from locally available organic materials such as logs, bundles of reeds or inflated animal skins, and which relied on the natural buoyancy of their individual components to float. It has been suggested that Homo floresiensis used bamboo rafts to reach Flores.

More sophisticated boats used hollow hulls to displace water and achieve buoyancy, an innovation that may have begun in Europe with the dugout canoe, or log boat. The Pesse canoe from the Netherlands is believed to be the world's oldest. A 118in (3m) long Scotch pine trunk hollowed out using stone tools, carbon dating indicates it was constructed during the early Mesolithic period between 8040BC and 7510BC. The discovery of the c.6,250 years old Dufuna Canoe in Nigeria suggests the basic dugout design was adopted independently by civilisations worldwide alongside other forms of watercraft, such as animal hides stretched over frames and woven baskets. These could have served multipurpose roles including fishing, hunting and cargo transport in rivers and along coastlines.

Bronze Age seafaring

The revolutionary discovery of metal working in about 3200BC marks the start of the transition from a Low to Median Technology phase in seafaring. The resulting diffusion of new tools manufactured from copper and bronze such as axes, adzes and saws allowed a dramatic increase in the range and sophistication of boat building techniques. Dug outs were limited in size and shape by the natural form of the tree trunk they were carved from, but timbers could now be efficiently split or cut into planks and composite hulls could be built up from numerous timbers trimmed to shape and attached using techniques such as sewing or pegging.

Among the earliest remains of plank-built vessels are the numerous well preserved examples found buried as funerary offerings in the deserts of Egypt. These include the Abydos boats, dated to the early first dynasty (2950–2775BC) and the famous Khufu Ship 1 (2600BC), one of two ceremonial barges found disassembled in sealed pits outside the enclosure wall of the Great Pyramid of Khufu at Giza. Constructed from 651 timber pieces and 667 fastening tenons, it is a large and graceful 143 feet (43m) long vessel, with collapsible

Fig 5.1: Khufu Ship 1, dated c.2600BC, preserved in the Giza Solar Boat Museum. (*Giles Richardson*)

deckhouse and canopy. The use of lines of papyrus reed stitching running transversely across the inside of the hull to lash together the planking suggests the boat builders were adapting techniques traditionally used for papyrus raft boats. However a new feature was the carving of mortise-and-tenon joints along the plank edges: square pegs that slotted into sockets cut in adjacent timbers and which added rigidity to the hull.

The earliest known group of seagoing craft are the partial remains of three boats discovered at North Ferriby in East Yorkshire, dating between 2000 and 1700BC, and the later and better preserved

Dover Boat dated to 1575–1520BC. These may represent the type of cross-channel trading vessel that brought the continental style axes and swords found in contemporary British settlements. Each could have transported about seven tonnes of cargo, and were either sailed or paddled by a crew of up to eighteen. They were built of oak planking stitched together with withies of yew along the length of their seams. The earliest known Mediterranean wreck, the Uluburun ship, found off the south west coast of Turkey, dates from the late Bronze Age c.1300BC, and carried a cargo of high status goods from up to ten different cultures from around the Eastern Mediterranean, including ten tonnes of copper ingots, one tonne of tin, raw glass, rare hard woods, ivory, and a wooden diptych, the world's oldest surviving book; an indication that the sea was now a highway for trading and gift exchange between elites.

Although there is no physical evidence for a mast or rigging on Bronze Age wrecks, depictions of boats with square sails on vases and wall paintings from 3100BC onwards make it clear sailing was well known in Egypt by this date. Sails appear in depictions in the Mediterranean from about 2000BC, normally as a square sail supported on a single mast at amidships. These were probably made from linen, reinforced at the corners with leather patches. The only example to survive from the ancient world is a fragment reused as a burial shroud in second century BC Egypt that still had a wooden brail ring attached. Such finds hint at the longer seagoing voyages powered by the wind that were now possible, and it is no coincidence that their appearance coincided with a dramatic increase in seafaring activity.

Shipping in the Archaic and Classical world

By the eight century BC a distinction was made by ancient writers between specialised warships and cargo ships. The Kyrenia wreck that sank c.300BC off Northern Cyprus with a cargo of 404 wine amphoras may have been a typical Greek merchant ship. Measuring 47ft (14m) by 14ft 6in (4.2m), a shapely 'wine glass' like hull profile with an extended keel would have balanced improved sailing performance with the maximum capacity for cargo. Unlike Bronze Age sewn plank boats, the builders only used mortise-and-tenons to fasten the joints between planking in the hull, which were now locked in place by small wooden pegs (treenails), hammered vertically through the hull, a technique possibly inspired by Phoenician ships. Copper nails now held frames to the planking.

Other innovations include the use of lead sheathing to protect against fouling and attack from wood-eating toredo worms.

The warships of Homer's *Odyssey* and *Iliad* used sails and single banks of thirty or fifty oarsmen. Double banks of oars appear in images of Phoenician ships by 700BC, and by 480BC triple banked triremes were the backbone of both Greek and Persian fleets at the battle of Salamis. Bronze rams may have been fitted from the late ninth century, first as simple spikes and by c.525BC as three horizontal blades. These were probably used to break opponent's oars, allowing enemy ships to be isolated and boarded by troops massed on deck. Popular images of triremes sinking enemy vessels through ramming may be exaggerated, but a recent find of a badly dented ram from the Egadi Islands, site of the First Punic War clash between Carthage and Rome in 241BC, suggests it had collided with another vessel head on. Oared warships reached their zenith in the Hellenistic navies where enormous vessels with five or six banks

Fig 5.2: The *Kyrenia Liberty*, a modern replica vessel designed to test the sailing characteristics of the ancient Kyenia Ship. (*Giles Richardson*)

of oarsman are recorded. When not in use, warships were stored in specially built ship sheds, found throughout the Mediterranean. The naval harbour at Carthage included ship sheds built on a circular artificial island. Harbour facilities such as Piraeus at Athens and Alexandria in Egypt were well developed, equipped with warehouses, quaysides, and man powered cranes to unload ships docked alongside. Outside the major population centres vessels were simply beached to unload cargoes.

From the mid second century BC Rome gradually conquered neighbouring civilisations and by the first century AD became the undisputed seafaring power in the Mediterranean. The technological advances at sea during the classical period were then stimulated by vastly increased demand for trade and transport between the city of Rome and its provinces. This is reflected in the emergence of merchant ships of much larger tonnage and carrying capacity. The La Madrague de Giens wreck from the first century BC was 130ft (40m) by 30ft (9m) and 15ft (4.5m) deep, and sank loaded with over 400 tonnes of wine amphorae – compared to just 25 tonnes on the Kyrenia wreck. Other ships, such as the Albenga from Italy could have been even bigger at 500–600 tonnes. These vessels supported their great weight through double layers of planking fastened by mortise-and-tenons, a feature that gave much greater hull strength than simply increasing the thickness of the planks. Merchant ship hulls also became fuller and flatter to increase capacity. A series of experimental vessels were built under the patronage of various emperors to transport eight ancient Egyptian obelisks to Rome, the largest weighing over 230 tonnes. Their designs are not well understood, but are likely to have been double-hulled barges that carried the obelisk underwater slung between the two hulls. Evidence of new equipment on Roman era wrecks includes pumps utilising the Archimedes screw, and leather valves. Larger Greek vessels could be equipped with multiple masts carrying square sails, but Roman vessels are the first to be depicted with a forward raked foremast, sometimes doubling as a crane for moving cargo, and with topsails.

As the volume of Roman shipping increased, so too did port facilities. Ostia, the old river port of Rome, was superseded under the emperor Claudius by new facilities at Portus that included an enormous sixty-nine hectare harbour basin protected by artificial breakwaters and a lighthouse mirroring the famous example at Alexandria. Here seagoing vessels transferred their cargoes to barges for transport upriver to the capital. When this provided insufficient shelter during storms in AD62 that destroyed numerous ships, it was further expanded under Trajan, who created a new thirty-nine hectare hexagonal inner harbour that gave direct access to multi-storey warehouses. Hydraulic concrete that could set hard underwater allowed convenient new harbours to be established around the empire on coasts not blessed with natural anchorages, such as Caesarea Maritima in Israel. Lighthouse provision was expanded: the Roman tower at Dover still stands today. The largest ports such as Rome, Alexandria and Constantinople probably served as entrepôts for long distance deepwater shipping routes, much like hub airports today, with networks of smaller regional and local harbours connecting to them by slower coastal sailing routes. Shipping costs for goods such as decorated ceramics were reduced by transporting them alongside bulk commodities such as grain and wine, which allowed merchants to reach otherwise uneconomical markets. A collection of Roman decorated pottery from Southern France found in the sea off Pudding Pan, Kent, was probably one such cargo.

Medieval seafaring

Roman and earlier plank ships were designed shell-first, i.e. the shape of the hull was determined by the planking during building, and the structural strength of the vessel came from that shell of planking, tightly held together by sewing or tenon fastenings. Framing was added later and simply reinforced the shell. This seems to have been the earliest boat building design used in all regions of the world. From the first century BC to the sixth century ad there are indications in the Mediterranean that this form of ship building was changing. Planking became less important structurally as the mortise-and-tenon fastenings became smaller, more widely spaced, and no longer locked. Rather than holding planks together, tenons now simply aligned them during construction. Instead the framing structure became the main source of structural integrity. In the new form of frame-first ship building, the frames were designed and erected first, and the planking fastened directly to the frames without the need to attach planks together. This became known as the carvel technique. The Serçe Limani I wreck, dated to 1025ad and found near the Island of Rhodes in Turkey, is the earliest example. The ship was a small two-masted merchant vessel, 45ft (15m) long with a boxy shape, capable of carrying thirty-five tonnes. All of the planking is nailed directly to the frames. Gaps between planks were caulked, and no fastenings were needed. Possibly induced by socioeconomic changes in Byzantine society, where cheap labour to craft intricate fastenings was no longer available, the switch to frame-first shipbuilding allowed ships to be built bigger, more quickly, and to be made more seaworthy. It also meant that hull shapes could be designed in advance of construction using frames as patterns, and successful designs could be copied and repeated. The science of marine architecture was effectively conceived at this point.

In Northern Europe, ship building developed differently. From the fourth century ad onwards, Scandinavia, the Baltic Coast of Germany and Poland, Britain, and Ireland, shared a boat building tradition that reached its peak during the Viking Age (ad800–1100). These vessels retained a shell-first design, but developed strong V-shaped hull forms formed from overlapping planking that was fastened with iron nails, known as the clinker technique. The Nordic method of construction produced a strong and light structure that could be easily adapted to build vessels of different size or function and which was well suited for Viking voyages. In the thirteenth and fourteenth centuries these designs evolved into larger medieval cargo vessels and warships such as cogs (a vessel with high sides and straight stem and sternpost) and hulcs (a vessel with high stem and stern) built using heavier timbers. The Bremen Cog wreck from AD1380 was 79ft (24m) long and could carry about 130 tonnes of cargo. It was equipped with both a windlass and a capstan for handling the rigging and anchors, among the earliest examples of these devices. Images from seals and coins show such vessels were the first to be equipped with superstructures at bow and stern (bow and stern castles), tops at the masthead for lookouts, and stern rudders moved by tillers from about AD1250.

Early modern shipping

From 1300, annual trading voyages between the Italian city states and Britain and the Low Countries allowed sailors to encounter both Mediterranean and Northern European ship building traditions. Cogs, with their large carrying capacity, seaworthy hull and stern rudder seem to have been adapted by Italian shipwrights who copied the design using the now established frame-first carvel techniques, and added lateen sails, naming them 'cocha'. These were then re-imported back to Northern Europe where they became known as 'carracks'. Henry V's warship *Grace Dieu*, completed in 1418, was an experimental English adaptation of the carrack, and was one of the largest ships built using this method. A 164ft (50m) long warship of over 1,400 tonnes, she used a frame-first construction but with three layers of overlapping planking in the traditional clinker style. Building such a large ship this way was expensive: seventeen tonnes of nails and one thousand beech trees were used. The ship also proved too expensive to maintain, and was abandoned and burnt in the River Hamble by 1439, where her remains lie alongside that of the *Holigost*, a smaller Spanish built carrack used by the Royal Navy until 1422.

By 1430 another southern European frame-first design, the 'caravela', probably based on Portuguese fishing boats, was imported, where it became known as the 'caravel'. The carrack and the caravel encouraged northern European merchant ship owners to move to frame-first ship building, becoming the designs that would be adapted to become the ships of choice for European explorers. For Columbus's fleet of 1492, the *Niña* and *Pinta* were caravels, and the *Santa Maria* a carrack. Later European sailors carried these designs to Arabia and India and perhaps China, where frame-first construction appears from the fourteenth century.

Frame-first construction also had a profound impact on naval warfare, as rows of gun ports could now be cut in the sides of a ship without weakening the hull, an impossibility in clinker-built ships. Ports protected from the sea by hinged lids also allowed heavy guns to be operated lower in a ship's hull without adversely affecting stability. The *Mary Rose*, Henry VIII's flagship launched in 1511, became the first ship to use purpose-built gun ports when she was refitted in 1536. Previously clinker built warships had mounted small guns on deck for anti-personnel use from about 1340, much as Greek triremes had carried archers and spearmen. Now large cannon could be used to sink enemy ships themselves, fundamentally changing naval tactics. By the mid seventeenth century large warships carried multiple rows of gun ports and were reinforced to protect against cannon fire, since gunnery at range was rapidly developing. The size of warships steadily increased to accommodate more guns with bigger calibres: *Sovereign of the Seas* of 1637 measured 1,522 tonnes and was the first to mount over 100 guns arranged over three decks. The *Victory* of 1737, carrying 100 larger guns, was 400 tonnes heavier, and by 1808, 120-gun warships of 2,600 tonnes and 197ft (60m) length such as HMS *Caledonia* were

Fig 5.3: The *Mary Rose* undergoing conservation in Portsmouth. The gun ports are clearly visible on the main and upper decks. (*Giles Richardson*)

being launched. In comparison, East Indiamen operated by the various European East India Companies were the largest merchant ships regularly built during the late eighteenth and early nineteenth centuries. One of the largest was the *Earl Talbot* launched in 1796 of 1,439 tonnes and 174ft (53m) length.

Larger ships created new technological challenges. Traditional hand pumps, using suction through leather valves to drain excess water from the lower hold, gradually gave way to more efficient endless chain pumps from the 1570s. Tillers, simple levers connected to the rudder and moved by multiple men, were replaced with geared steering wheels around 1700. Copper sheathing to protect ships hulls was tested on the frigate HMS *Alarm* in 1761, and introduced for general naval use from 1783, a change that required all iron fastenings below the waterline to be replaced with copper bolts to prevent electrolytic corrosion. The science of naval architecture was significantly advanced with the beginning of the study of the mathematical principles underlying stability and performance. Draughtsmanship of hull line plans allowed successful designs of both warships and merchant vessels to be shared through published treatises on shipbuilding, reducing reliance on the skill of the individual shipbuilder.

Shipping in the Industrial Age

As the Royal Navy expanded through the eighteenth century, the physical limit of all-wooden ship size was reached and domestic timber supplies for shipbuilding were gradually exhausted. By the Napoleonic Wars (1793–1815) alternative solutions were tested, including the use of diagonal iron straps (riders) to strengthen the lower hull planking of HMS *Tremendous*, launched 1811. The man responsible for this innovation, Sir Robert Seppings, was appointed Surveyor of the Navy in 1813, and over the next decade he proceeded to introduce a radical new system of warship construction based on new manufacturing technologies. A lattice of overlapping diagonal framing, first made from timber and later from iron, replaced traditional vertical frames as the structural skeleton of the ship. Elsewhere iron knees replaced the large and increasingly costly wooden pieces that joined deck beams to the frames, while heavily reinforced circular sterns replaced vulnerable window filled transom sterns. The new stronger hulls could support heavier and more powerful guns, provided better protection against enemy fire, and were more able to withstand the stresses of bad weather at sea without damage, dramatically extending a ship's useful service life. All of these innovations can be seen in HMS *Unicorn*, a Seppings designed frigate launched in 1824 and now preserved in Dundee as a unique survivor from the transitional period between wooden and iron ships. The new system of construction also allowed the largest class of three-deck warships to be built. HMS *Duke of Wellington* launched in 1852 was 240ft (73m) long and mounted 131 guns. The 1858 Mersey-class steam powered frigates HMS *Mersey* and *Orlando* were the longest wooden warships ever built, at 335ft (102m) length overall. However the extreme length combined with the weight of the engines put enormous strains on their hulls, causing structural failures even with iron reinforcements, and both were scrapped by 1875. The short careers of these ships indicated that they had already exceeded the practical limits for the size of wooden hulls.

The first steam-driven ship in Royal Navy service was the paddle steamer HMS *Comet*, launched in 1822 and employed as a tug for towing sailing vessels in the Thames and Medway. However the Navy was much slower to adopt steam-powered

Fig 5.4: HMS *Duke of Wellington* 1852. The most powerful warship in the world when launched, her design represented the ultimate development of the multi-decked ship of the line that had dominated naval warfare for over 200 years, yet was obsolete within a decade. (*Photograph by Symonds & Co, c.1880*)

fighting ships, largely due to concern for the vulnerability of the engines and paddlewheels exposed to enemy fire. The success of the first screw-driven ship SS *Archimedes* in 1840 convinced the Admiralty to conduct further tests of the technology. In March 1845 trials investigating the relative merits of screw and paddle propulsion culminated in the famous tug-of-war contest between the experimental screw sloop HMS *Rattler* and the paddle frigate HMS *Alecto*, which *Rattler* conclusively won by towing *Alecto* astern at a speed of 2.8 knots. From this point on all newly built Royal Navy vessels would be screw-driven, although engines were still considered ancillary to sail power, so battleships retained full sailing rigs well into the early 1880s.

Steam was also slow to supplant sails for most forms of long-distance trade. Up to 1870 three-masted clipper ships, constructed from wood, iron or a composite of both, offered faster transport between Europe and the Far East or Australia than contemporary steamships. However the opening of the Suez Canal in 1869 shortened these routes by several thousand miles, and combined with improved engine technology, tipped the balance towards steamships that came to gradually dominate all forms of maritime transport. Sailing merchant ships continued to offer economic advantages over powered vessels, such as lower running costs, smaller crews and no consumption of fuel. Large four or five-masted Windjammers of 2,000 to 8,000 tonnes with iron or steel hulls continued to be profitable in general cargo trade until the Second World War, when they were finally replaced by cargo tramp steamers.

By 1858 Britain had built or converted thirty-two wooden ships-of-the-line to steam power. However results obtained by new rifled guns firing explosive shells showed how vulnerable such vessels had become to new armament technology. Meanwhile France's launching in the following year of the ironclad wooden-hulled frigate *Gloire*, the first ocean-going armoured warship, upset the balance of power by rendering all unarmoured ships obsolete. In response the Admiralty ordered the building of the 420ft (128m) long HMS *Warrior* and *Black Prince*, the first large warships built with completely iron hulls. The *Warrior*-class design successfully

Fig 5.5: HMS *Warrior* 1860. The twin funnels for her steam engine are clearly visible between the traditional ship rigged masts. (*Photograph by West & Son, Gosport, c.1870*)

merged the proven technologies of screw-driven steam power, iron hulls and rifled guns with new wrought iron armour to produce the largest, fastest and most powerful warships ever built. The armour consisted of a 213ft (65m) long citadel amidships that protected the main battery of twenty-six guns with 4.5in (11.4cm) thick wrought-iron plating and an additional 18in (45.7cm) of teak backing.

By 1862, navies across Europe had adopted ironclads. Britain and France had sixteen each, either completed or building. However the rapid evolution of ironclad development, for which the Warrior-class was partly responsible, meant that the two vessels were effectively obsolete within a decade of their launch. Instead of mounting as many guns as possible in a broadside battery, designers concentrated on smaller numbers of larger calibre guns to punch through enemy vessels' armour. In 1873 the Royal Navy commissioned its first class of capital ship to be designed without masts, the turret ships HMS *Devastation* and HMS *Thunderer*. Without the hindrance of sailing gear, the main armament of four powerful 12-inch guns could be mounted in two revolving turrets on the main deck that allowed a 280 degree arc of fire. This has been described as the most radical design of the nineteenth century, marking the beginning of the end of the sailing warship. A comparison of the profiles of *Duke of Wellington*, *Warrior* and *Devastation* illustrates just how rapidly naval design had advanced in the two decades during which these three vessels entered service. Warship firepower, armour and speed continued to develop. By 1906 HMS *Dreadnought* was 525ft (160m) long and over 18,000 tonnes displacement. Her uniform heavy armament of ten 12-inch guns mounted in five turrets, steel armour and unprecedented speed of 21 knots so thoroughly eclipsed earlier designs that subsequent battleships became generically known as 'dreadnoughts' and preceding battleships of the 1880s and 1890s disparaged as 'pre-dreadnoughts'.

The change from wooden to iron ships caused a shift in the focus of the ship building industry that had for centuries been concentrated in the South of England close to the source of timber supplies. Instead the iron ship building industry moved north to the new industrial centres of Glasgow, Liverpool, and the North East that could supply coal, iron and skilled workforces to harness the new technology. The adoption of steam propulsion and rapidly

Fig 5.6: HMS *Devastation* 1871. The 12-inch revolving gun turrets sit either side of the raised superstructure. (*Photograph by Symonds & Co.*)

increasing size of warships also led to large-scale changes in dockyards, which were far too small to cope with this revolutionary change in ship building and maintenance. In 1843 reclamation work began in Portsmouth harbour to create a new seven-acre basin with dedicated steam workshops and dry docks alongside. However the introduction of the Warrior-class meant that these new amenities had to be rebuilt and expanded almost as soon as they were finished. From 1867 to 1876 the 'Great Extension', added a complex of three new large interconnected basins, three dry docks and 530yd (4,600m) of wharfing. Some 582,700,000 cu.ft (16,500,000cu.m) of spoil was excavated and used to dramatically expand the nearby Whale Island. By the end of the century still further expansion was needed in order to keep pace with the increasing size of battleships, and the intervening walls between the basins were removed in 1912 to create a single large basin of fifty acres which remains in use to the present.

In conclusion it is clear that the history of shipping technology is a study of both rapid and slow changes. For 5,000 years ship building innovation was focused on the ability to harness the power of the wind and the limitations of timber as a raw material. Throughout this Medium Technology phase, seafaring underwent continuous development, so by the beginning of the nineteenth century wooden sailing vessels had reached the pinnacle of technical achievement. In contrast, the shock of industrialisation and the introduction of iron and steam power led to enormous change over a single century, and can certainly be described as a high technology revolution. However the basic aims of ship builders remained the same across time: to construct faster, safer and bigger ships capable of carrying more cargo, or naval vessels capable of out-fighting or out-running the enemy. Further twentieth century innovations of welded steel Post-Panamax super tankers, aircraft carriers or nuclear powered submarines ultimately also conform to these aims. As such, in the modern world, where a remarkable 90 per cent of world trade still travels by sea, and maritime security remains a strategic necessity, mankind's use of technology in the maritime world is as important as ever.

Further reading

Bass, George F: *Beneath The Seven Seas: Adventures with the Institute of Nautical Archaeology*, (London, 2005).

Delgado, James P (ed.): *The British Museum Encyclopaedia of Underwater and Maritime Archaeology*, (London, 1997).

McGrail, Seán: *Boats of the World: From the Stone Age to Medieval Times*, (Oxford, 2001).

Parkes, Oscar: *British Battleships*, (London, 1956).

CHAPTER 6

Transport II – The Steam Revolution on Land and Sea

Angus Buchanan

The Romans built a magnificent network of roads to serve the strategic requirements of their armies in governing a huge empire, and Britain acquired an extension of this transport system when it was assimilated into the Roman Empire. The roads were carefully surveyed and well designed, taking the most direct routes between the major administrative centres. They were given a firmly compacted and convex surface, frequently paved, with ditches on both sides to assist drainage. These roads survived for many centuries after the Western Empire had broken up into

Fig 6.1: Menai Straits Bridge. The splendid suspension bridge completed in 1826 by Thomas Telford embellishes the fine landscape of Menai and Snowdonia. (*Angus Buchanan*)

Fig 6.2: The Iron Bridge. The first large cast-iron bridge was erected by Abraham Darby III in 1777 over the River Severn at the township, which came to be called 'Ironbridge'. (*Angus Buchanan*)

a large number of rival states, providing the best routes for land transport with little maintenance except that supplied by local initiatives, which tended to be uneven and spasmodic. The result was that by the beginning of the eighteenth century the standard of British roads was notoriously bad, with a few services available on horseback but best avoided altogether if alternative means existed by navigable rivers or coastal traffic.

New roads and canals

The needs of a country with increasing industrial activity eventually determined that something must be done to improve these conditions. The military roads to pacify the Highlands of Scotland built under the direction of General Wade had demonstrated the strategic value of better roads by the 1730s, and Turnpike Trusts were encouraged by Acts of Parliament to undertake the upkeep of stretches of road in return for the privilege of collecting tolls from users of the King's Highway. These Trusts provided significant improvements, but in a rather patchy and disjointed fashion. They showed what could be achieved by private enterprise with official approval, and gave a generation of road builders such as Thomas Telford and John Loudon McAdam the opportunity to practise their skills and to acquire a reputation as road engineers. By the 1830s, therefore, Britain had been equipped with a network of decent main roads over which mail coaches and goods traffic could move around the country with a greatly improved facility.

The same sort of state-supported private enterprise had meanwhile been stimulated into constructing the first artificial waterways in Britain. These were already well established in France and elsewhere in Europe, notably the Canal du Midi in southern France, commissioned by Colbert in 1666 to encourage the wheat trade and built by Pierre-Paul Riquet and completed in 1681: it was 150 miles (241km) long, with pound locks and

Fig 6.3: Conwy Bridges. The Telford suspension bridge stands in front of Robert Stephenson's tubular bridge to cross the river at the foot of Conwy Castle. (*Angus Buchanan*)

aqueducts, and by linking with other canals became the Canal des Deux Mers, linking the Atlantic with the Mediterranean. Canals were introduced to Britain in 1759 by the Duke of Bridgewater when he commissioned the engineer James Brindley to build a canal to transport coal from his estates at Worsley five miles into Manchester. The value of this canal in shifting a bulky commodity safely and cheaply was quickly recognised by other entrepreneurs in the Midlands, who raised capital by forming private companies authorised by Act of Parliament to build a canal over a carefully specified route, and to acquire the land for this purpose by compulsory purchase if necessary. Within a few decades, a network of canals had been created to serve the needs of heavy industries in Britain, getting coal and raw materials to their factories and conveying their manufactured goods safely to market. These early canals were narrow and avoided gradients as far as possible, but where there was no alternative tackled them with pound locks, canal lifts, aqueducts and tunnels. They provided a great stimulus

Fig 6.4: Pont Cysylte Aqueduct. Thomas Telford's multi-arched aqueduct crosses the River Dee in a cast iron trough 120 feet (37m) above the river and is still in service. (*Angus Buchanan*)

to manufacturing industries in parts of the country previously considered too land-locked to permit industrial development.

With growing experience of canals and increased engineering confidence, the canals grew in size and complexity, promoting cross-country routes such as John Rennie's Kennet & Avon Canal linking Bristol to London, and John Smeaton's Forth & Clyde Canal across the central lowlands of Scotland. They also led to strategic canals such as Thomas Telford's Caledonian Canal through the Great Glen in Scotland, which in theory avoided the need for naval vessels to sail round the north of Britain. By the time it was complete, however, in 1822, the strategic need for such a short cut had faded, and for all its magnificent engineering the whole canal system was on the point of becoming virtually obsolete. The reason for this was that the steam engine had already begun to provide alternative transport systems in the shape of the railways and the steam ship, which promised to deliver more quickly and reliably than either the canals or the turnpike roads.

The Railway Age

The value of designated tracks of rails in the form of iron plateways or tramways to carry commodities in wagons was first exploited by British colliery owners in the North East coalfield to carry coal from their mines to wharves on the Tyne, the Wear, and the Tees, from which it could be shipped to London and other markets, where it was welcomed as 'sea-coal'. It was here, also, that the ability of the steam locomotive as invented by William Murdoch and Richard Trevithick was first applied to the task of pulling heavy 'trains' of wagons over prepared stretches of the track. The key person in promoting this transport system was a colliery engine-man, George Stephenson, who developed remarkably successful steam locomotives for this purpose, and quickly realised their potential for securing a much wider market for his coal. Stephenson's vision became that of constructing a network of railways serving markets all over the country. As a first step towards this objective he persuaded a group of entrepreneurs to establish a company for which he built the Stockton & Darlington Railway, opened in 1825, and equipped it with his *Locomotive No.1* as its first engine.

The idea quickly caught on, and another group of merchants in Lancashire formed the Liverpool & Manchester Railway (L&MR) to link the two towns and commissioned George Stephenson to build it for them. Although not a professional civil engineer, he overcame

Fig 6.5: *Mallard* at speed. This fine LNER locomotive held the world speed record for many years. (*J S-M*)

some serious problems in laying a level and stable track, and the railway was officially opened in 1830, becoming the first main-line railway in the world running a regular scheduled service of goods and passenger trains. Stephenson had established his son Robert as a mechanical engineer, and the young man had shown exceptional talent in designing a new locomotive for the railway, which had won the competition held at Rainshill, near Liverpool, in September 1829 to determine which power unit to adopt. It was this engine, *Rocket*, with a multi-tube boiler and steam blast and diagonally placed cylinders driving directly onto the front two wheels, which thus set the style for the L&MR. The Stephensons immediately capitalised on their success by improving the design and by placing the engine cylinders horizontally under the boiler, thus producing in *Planet* the arrangement which became the standard design for the great expansion of the steam railway system that sprang from it.

Thus equipped with sound track and excellent locomotives, the Railway Age entered its century-long dominance of land transport. Beginning in Britain in the 1830s, the main rail routes out of London were all established by the early 1840s, and the success of these early ventures engendered a vogue of railway speculation which became known as the 'Railway Mania', until over-enthusiasm and financial chicanery precipitated a slump in confidence in 1848, and thereafter expansion resumed at a more modest rate. The technical success of the railways, however, in carrying an increasing volume of goods and passengers faster than had previously been conceived possible, had a profound effect on the economic life of the nation. Coal mining and the heavy metal industries received a vigorous stimulus, and engineering expanded to cope with the demand for both civil and mechanical engineering, while people were encouraged to travel further and more frequently than they had previously managed to do. Cheap travel early and late in the day enabled workers to live some distance from their place of employment, stimulating urban growth in new suburbs. When the Great Exhibition was held at the Crystal Palace in Hyde Park in 1851, the railways acquired a popular new function in arranging special excursion trains to bring visitors up to London. Cultural and sporting activities became more accessible, and publications for the travelling public flourished. The railways in fact produced a social as well as a technological revolution.

When I K Brunel, the engineer of the Great Western Railway from London to Bristol, died at the age of fifty-three in September1859, to be followed within the next year by the death of Robert Stephenson and Joseph Locke, *The Times* saluted them as the 'Railway Triumvirate', the railway engineers who had transformed the life of the nation. There was a degree of hyperbole in this praise, as many other engineers, entrepreneurs, and public-spirited persons had contributed to the achievement, but they were certainly the outstanding figures – after George Stephenson, who effectively retired from railway construction in the early 1830s and died in 1848 – in the first generation of British railway building. Between them they had master minded the construction of a national network of major routes reaching all the main centres of population, and by their vigorous but friendly rivalry they had initiated striking advances in the quality of railway services. They also determined the national gauge of the track, with Brunel having to concede the victory to the narrow or 'standard' gauge of 4ft 8½in (1.43m), despite the operating advantages of the broad gauge of 7ft (2.13m) on which he built the Great Western Railway, designed to carry high-speed passenger traffic. This was a daring innovation, but Stephenson and Locke had already established many miles of

Fig 6.6: The graveyard of steam locomotives. When steam locomotives were phased out by British Railways in the 1960s, many of them were parked in these sidings at Barry Dock in South Wales from which preservation societies subsequently salvaged them for their railways. (*Angus Buchanan*)

standard gauge and when the inconvenience of changing between track of different gauges became apparent, it was simpler for the broad gauge to be narrowed rather than *vice versa*. The GWR carried on with the broad gauge until the 1890s, but then converted to standard gauge.

By the 1860s, virtually every available gap in the British network had been filled in and, following British leadership, the rest of the world had accepted the dominance of the steam railway in land transport. As early as the 1830s all the members of the 'Triumvirate' had been receiving commissions to build railways in Europe and North America, and before long British engineers had gone out to Africa, India, Australia, and South America, taking British skills and experience to construct railway systems wherever a need could be found for them. The British engineering industry provided steam locomotives and rolling stock for most of these ventures, but gradually the recipients developed their own expertise and the British participation dwindled. Nevertheless, the dynamic imparted to the British economy by the steam railway was exported all over the world, so that it may be regarded as an undoubted technological triumph. When the locomotive steam engine was eventually overtaken by the internal combustion engine and electric traction, British railways held on to their reliance on steam engines as long as it was economically feasible to do so. However, the savings in arduous labour and the smoothness of operation when diesel and electric power became available had to be recognised, so that steam power disappeared from main-line British train running in the 1960s.

Steam at sea

While the steam locomotive was transforming land transport, a parallel revolution was occurring at sea. When Nelson received his fatal wound on HMS *Victory* at the Battle of Trafalgar in 1805 British naval supremacy was secured for another hundred years. *Victory* herself was an old ship, representing the end of a fine tradition of naval warships of multi-gunned, wind-powered, wooden ships going back at least to the *Golden Hind* of Sir Francis Drake in the sixteenth century. A hundred years after Trafalgar, however, this type of ship had become completely obsolete, to be replaced by steel-hulled vessels carrying massive guns capable of firing in any direction, and achieving high speeds from the power of steam turbines – the ships modelled on HMS *Dreadnought*, which was under construction in 1905. The first experiments in steam propulsion for ships had taken place on Scottish canals at the end of the eighteenth century, using Watt-style beam engines to drive paddle wheels, and by 1812 the first steam-powered ferry boat *Comet* was in service on the Clyde. Small ships of this type were found useful for passenger ferries and river-tug duties, and many such vessels, usually with the overhead beam replaced by a low-level side-lever in order to improve their balance, were introduced to such duties in ports and river estuaries during the 1820s.

Early steam ships operated under a serious constraint because they needed frequently to replenish their coal to fuel the steam engine. It seemed at this stage that there was no long-distance prospect for steam propulsion at sea because, by scaling up the coal required for a large steam ship, it would need to occupy a prohibitive proportion of the available volume for a long voyage, thus making it commercially useless. However, in 1838 I K Brunel, anxious to extend his Great Western Railway from Bristol to New York, recognised that it was the cross-section of the ship rather than its total volume that controlled the fuel requirements to propel the ship through the water, and persuaded a group of Bristol merchants to invest in the construction of a sleek ship, designed by himself to cross the Atlantic Ocean. This was the steam ship *Great Western*, a substantial wooden-hulled vessel with paddle wheels driven by a side-lever steam engine, which made its first crossing to New York in fifteen days, with plenty of coal still unused in its bunkers.

The viability of trans-Atlantic steam transport thus having been dramatically demonstrated, other operators were quick to follow. The Bristol merchants were keen to maintain their advantage over their rivals, and commissioned Brunel to build them a sister-ship to the *Great Western* which would enable them to run a regular shuttle service between Bristol and New York and thus secure the valuable mail contract. However, Brunel's vision had expanded and the ship he designed and built was not only much larger than its predecessor, but it incorporated two striking innovations: it was the first large ship built with an iron hull, and it scrapped paddle wheels in favour of screw propulsion. These novelties delayed its construction so that Bristol lost the mail contract, but

Fig 6.7: SS *Great Britain*, Brunel's second steam ship, ended her long working life as a hulk in the Falkland Islands. She was recovered and brought home to Bristol in 1970. (*Angus Buchanan*)

the SS *Great Britain* when at last launched and at sea in 1844, fully justified the acumen of its designer. It had a hugely successful commercial career, being adopted for the long-distance Australian run before eventually being abandoned in the Falkland Islands – only to be triumphantly brought home to Bristol in 1970, where its rusting hull has been repaired and the ship restored to resemble closely its original condition and to provide an exceptionally vivid experience for visitors and tourists.

The SS *Great Britain* was too large to work conveniently out of the cramped city docks which then constituted the port of Bristol, so that it was transferred for most of its working life to sail from Liverpool. In designing his third ship, the SS *Great Eastern*, Brunel realised that it would be far too large to be accommodated in Bristol and chose to build it on the Thames, at Millwall on the Isle of Dogs, and even here it had to be launched sideways into the river – an arrangement that caused Brunel great anxiety. Such anxiety, however, accentuated many other problems encountered in the

Fig 6.8: The SS *Great Britain* now occupies the dry dock in Bristol where she has been gloriously restored to resemble her original condition. (*Angus Buchanan*)

construction of this gargantuan vessel – the largest ship to be built until the last decade of the nineteenth century. It was so big because Brunel's vision was of a ship sufficiently large to carry its own fuel for a journey to the Far East and back. Its hull was immensely strong, being double skinned and compartmentalised, all in the best quality wrought iron. It had two sets of steam engines – one with a horizontal cylinder driving a screw propeller, and the other an oscillating engine, with its two cylinders swinging up and down as they drove two huge paddle wheels.

Brunel recognised that not even he could master mind every detail in the construction of this huge ship, and he willingly entered into a partnership with an outstanding ship-builder, John Scott Russell, in whose shipyard it was constructed. Russell had made his reputation with his 'wave-line' theory, by which the lines of a large ship could be designed to ensure maximum efficiency in any sea conditions, and this was employed in the *Great Eastern*. Unfortunately, Brunel and Russell had very different temperaments and different ways of working, and the relationship between them became increasingly bitter, which greatly complicated the problems of building and launching the ship. This was achieved by 1859, but the anxiety contributed to Brunel's premature death in September of that year, while the ship was making its maiden voyage – during which a boiler-casing exploded, sending one of the ship's five funnels flying into the air and killing several stokers.

The SS *Great Eastern* did not repeat the commercial success of its two predecessors. It survived a series of misfortunes and performed a useful service in laying trans-oceanic telegraph cables, but Brunel had in fact misjudged the amount of traffic available to justify its huge expense, and it was only at the end of the century that sufficient business had been generated to make it worthwhile building such a big vessel again. Moreover, almost as soon as it was launched its engines became virtually obsolete as designers began to experiment with types of compound engines working with two or more cylinders arranged in series at reducing pressures. By the 1870s, forms of triple compounds, with three vertical cylinders placed along the keel of the ship, became normal practice in both the merchant marine and naval services, and remained such until steam turbines and internal combustion engines began to take over at the beginning of the twentieth century. Then the age of the great steam liners began with the turbine-powered RMS *Mauretania* in 1906, by which time a vigorous demand for two-way passenger traffic across the Atlantic had developed and continued – with the intervention of two World Wars – until the 1950s, when air-travel was becoming easily available.

Meanwhile, strong national rivalry had promoted sharp competition between the British Royal Navy and the navies of other powers, particularly France and Germany, with Italy, Russia, the USA, and Japan joining in by the end of the nineteenth century. The introduction of high explosives in place of gunpowder led to the development of large guns placed in rotating turrets so that they could cover all directions, instead of the banks of many cannons such as those of HMS *Victory*. Shells with increased power of penetration emphasised the need for ever-greater protection of the most vulnerable parts of the ship, so that hulls were designed with thick wrought iron plates, and subsequently of mild steel as this stronger and lighter metal became available in bulk. Of great importance in the transformation of naval vessels was the power unit, so that the navies were prompt to adopt the steam turbine, preferably with steam from oil-fired boilers, in order to acquire greater speeds and manoeuvrability. So emerged the *Dreadnought* combination of qualities, in time to equip the navies of the main combatants in the First World War.

There were important subsidiary innovations which kept the designers of rival navies alert, such as the torpedo – a powerful under water shell driven by its own power unit and guidance system – and the submarine, which could launch such weapons undetected from underwater. Then depth charges were introduced to destroy submarines under water. As soon as aeroplanes became feasible as adjuncts to naval warfare, attempts were made to use them for reconnaissance, and then for dropping bombs or launching torpedoes against capital ships. They made only marginal impact in the First World War, but by the Second World War aeroplanes had become weapons capable of destroying even the largest ships, as the Royal Navy learnt to its cost in 1942 when it lost two capital ships in the approaches to Singapore when attacked by the invading Japanese aeroplanes. Such disasters caused a hurried reappraisal of the role of large naval vessels in conflict and led to an emphasis on aircraft carriers and smaller ships grouped in battle fleets to protect the carrier and to destroy attacking aircraft. The age of the large battle cruisers was as effectively as over as that of the luxury passenger liners. In future, the big ships tended to be either oil tankers or cruise liners, driven by diesel engines.

New life for roads

While diesel engines and electric motors heralded the decline in the use of steam power on railways and at sea, they brought new vitality to road transport, which underwent tremendous expansion all over the world in the twentieth century. Although road traffic had declined in the 1830s with the initial success of the railways, sufficient local traffic had been generated to justify the continued maintenance of the main roads, many of which had been taken under the control of national or local government, so that they remained available for experiment with alternatives to horse-driven vehicles. The first of these was with steam power, with which there were several promising initiatives in the 1820s and 1830s. A succession of steam carriages appeared on British roads, designed by Sir Goldsworthy Gurney and other operators, which could have developed into a viable means of road transport but instead encountered so much hostility from other road users and the railways that they were frustrated by prohibitive legislation, such as the law requiring any mechanically propelled vehicle on the roads to be preceded by a man carrying a red flag. So the only steam vehicles to have any commercial success were steam traction engines for transporting heavy loads and steam rollers for road maintenance use, while bulk transport together with passenger and mails services passed almost completely to the railways for the rest of the nineteenth century.

The revival of interest in mechanical means of road transport appeared in the form of the bicycle. The discovery of the gyroscopic stability that could be achieved by balancing on two wheels led to a plethora of designs for bicycles in the middle of the nineteenth century, which, by the 1880s, had evolved into the so-called 'safety bicycle' with two equal sized, wire-spoked, wheels mounted on ball-bearings, and the rear wheel driven by pedals and a chain from a central axle. The addition of pneumatic tyres and a diamond-shaped frame of tubular steel then completed the archetypical bicycle design, which became the standard for the next century. In this form the bicycle met the rising consumer demand for a simple and convenient mode of transport, and was of inestimable social value in opening up a wider field of travel, particularly for women and young people. It also gave a significant boost to the engineering industry, which responded to the need for bicycles and the new components that they required.

Engineers soon applied themselves to the possibility of applying mechanical power to the bicycle, and in 1885 two German engineers, Gottlieb Daimler and Karl Benz, working independently, succeeded in adapting an internal combustion engine for this purpose. As long as this type of engine relied on coal gas as a fuel, it was as inappropriate as a steam engine for this sort of use because of its dependence on a local gasworks for its supply of fuel. But the development of oil-based fuels, and especially petrol, made it possible to build a compact engine carrying its fuel in a small tank, and it was in this form that Daimler managed to attach one to a bicycle to drive the first 'motor-bike', and Benz to do so with a larger three-wheeled cab which thus became the first 'motor-car' or automobile. These inventions, incorporating many of the features of the bicycle, were quickly taken up by other engineers, especially in Germany and France, although British engineers were slower off the mark as they were held back both by their traditional dependence on the steam engine and by the restrictions imposed by the 'Red Flag' Act. The repeal of this legislation in 1896 encouraged many British engineers and manufacturers to respond to the rapidly rising public demand for motor cars. Firms like Rover, established to manufacture bicycles, quickly switched to motor cars, and they were soon joined by Lanchester and Rolls-Royce. In Germany, Daimler and Benz had many followers, and in France the names of Peugeot, Panhard-Lavossor and Renault joined the engineering talent drawn into the new industry.

In the United States, Henry Ford made his fortune by producing motorcars that were available at a cost within the reach of the burgeoning artisan and middle class society that became a feature of all Western nations in the twentieth century. Ford perceived the huge market that was available if a sufficiently cheap but efficient vehicle could be produced, and proceeded to build up an industry capable of doing so. He achieved this by adopting a style of motorcar as a basic standard, and leaving any refinement or decoration to the taste of his consumer, who could have any colour he or she wanted 'providing it was black'. Such standardisation was made possible by devising a process of 'mass-production' that consisted of ruthlessly repetitive operations arranged round a moving conveyor belt on which a continuous series of vehicles was assembled, at the end of which the vehicles could be 'rolled off' to the waiting customers. The whole process followed the principles of 'work study' as devised by Ford's friend, F W Taylor. The workforce was hard driven but reasonably well paid, as Ford recognised that they and their families were his potential customers, but he resisted any attempt at unionisation. He set up his production lines at Dearborn, Michigan, in 1903, and in 1908 he began manufacturing his Model T, the 'Tin Lizzie', the famous vehicle which became a worldwide success, and by the time the production line was closed in 1927 some fifteen million vehicles had been made.

Like the steam engine before it, the internal-combustion engine underwent a long process of development. The new science of thermodynamics had suggested that the fuel would burn most effectively if it was first compressed in the working cylinder, and by arranging his engine with four cylinders acting sequentially – injection, compression, ignition and exhaust – on the same drive-shaft, the German engineer N A Otto demonstrated a smooth-running performance in 1876 – the 'Otto-cycle' – which was ideal for the engines used to propel motorcars. His compatriot Rudolf Diesel took this perception further in 1892 by proving that very high compression of the fuel would induce self-ignition, so that the Diesel engine became a popular method of powering heavy engines such as those used for tractors, omnibuses, and ships. There were many later developments of the internal-combustion engine throughout the twentieth century, particularly the gas turbine invented by Frank Whittle and contemporaries in Germany and elsewhere, working in ignorance of each other in the race to make a suitable engine for high-speed aircraft in the Second World War. Whittle's first experimental engine ran in 1937, and the two-engined Meteor jet plane came into service at the end of the war. Such engines subsequently became the mainstay of the air-fleets of all the nations, for both civil and military aircraft (see Chapter 7).

The world impact of the Transport Revolution

In its first and most ubiquitous application to transport in the form of the automobile, the internal-combustion engine has had a tremendous impact on society throughout the world in the twentieth century. For one thing, it broke the railway monopoly of passenger transport, providing a mode of travelling which was more convenient to use and more wide-ranging because its vehicles could be personally owned and directed to any destination accessible by road. Then the roads themselves underwent a dramatic revival of fortune, as they began to be preferred to railways for the conveyance of all sorts of commodities as well as people, and the construction of new arterial roads and motor-ways became an indication of national vitality, largely eclipsing the railways, although these remained indispensable for long-distance traffic as well as for urban commuter travel. Once in place, moreover, a good network of high-class roads has had a dramatic effect on the growth of towns and conurbations, creating a huge volume of traffic and dominating personal life in many aspects, not always wholesome, as in the creation of urban pollution. For better or worse, however, we have to live with the motor vehicle in all its forms, and it has come to support a massive industry in producing and servicing its needs, and thus to make a giant contribution to the national economy of modern states.

The role of steam in both marine and naval uses, as well as those in land transport, after a century of dominance, has thus become much diminished in modern means of transport. In its place electricity and the internal combustion engine have become the leading sources of power on land and sea and in the air and space. It is thus a mistake to consider our era as that of a 'post-industrial society', when we can summon power for lighting and heating at the press of a button, and move with ease between distant countries in automobiles and aeroplanes. We should rather consider what makes our era so different and distinctive from all those that have gone before, and if we do that our answer would probably mention the computer, the mobile phone, and the power of instantaneous communication with people all over the world, our relatively comfortable standard of living, our freedom from plagues and most infectious diseases, and the television, radio, and other sophisticated electronic devices which bring us armchair entertainment. Rather than living in a 'post-industrial' society, we live in a society in which enormous power has been harnessed to sustain our quality of life. If we can recognise that fact we can begin to hope of preserving it and extending it. Perhaps it deserves to be called the Age of the Electronic Revolution, succeeding both that of steam power and that of the internal combustion engine. Far from moving beyond industry, we should be seeking to understand where our phase of industrialisation is leading us.

Further reading

Brown, David: *Warrior to Dreadnought: Warship Development 1860–1905,* (London, 2004).
Buchanan, Angus: *Brunel: The Life and Times of Isambard Kingdom Brunel,* (London, 2002).
Hadfield, Charles: *The Canal Age*, (David & Charles, Newton Abbot, 1968).
Rolt, L T C: *Victorian Engineering,* (Routledge, London, 1961).
Simmons, Jack: *The Railways of Britain,* (Routledge, London, 1961).
Skempton, A W: *John Smeaton,* (Thomas Telford Ltd, London, 1981).

Chapter 7

Transport III – A Brief History of Aeronautics

David Ashford

This chapter starts with a brief narrative summary of the history of aeronautics. This is followed by a more detailed consideration of two particular aspects – why the aeroplane was invented when it was, and the influence of government. Finally, there is a brief consideration of safety and the influences of science and prizes.

Human interest in flight has a long history, birds providing inspiring exemplars. The legend of Icarus and Daedelus, and the sketches of Leonardo da Vinci are probably the best-known expressions of this interest. The Chinese may have built man-lifting kites in the seventh century AD. But, until 1783, the human race was in effect earthbound. Then the Montgolfier Brothers invented the man-carrying balloon. For the next 100 years, balloons were the only way to fly. They were used for niche purposes such as daredevil showmanship (charging a fee to enter enclosures), atmospheric research, artillery spotting, and carrying passengers on air experience flights.

During the nineteenth century several pioneers attempted to build aeroplanes in an attempt to overcome the main drawback of balloons, which is that they cannot fly into wind. The first sketch of a recognisable aeroplane was produced by Sir George Cayley in 1799, and he can justly be called the 'conceptual father of the aeroplane'. He identified the four forces that acted on a flying machine: weight, lift, drag, and thrust. He was the first to describe an aeroplane with the essential features of a light structure, fixed wings (i.e., not flapping like birds), a cambered wing section, and separate systems for lift, propulsion and control.

Despite Cayley's work and the efforts of others such as Stringfellow using models (1848), it was not until 1891 that consistent and controlled flight was demonstrated by Otto Lilienthal with his first glider. He became the first person to make well-documented, repeated, successful gliding flights. He followed an experimental approach established previously by Cayley. Newspapers and magazines published photographs of Lilienthal gliding, favourably influencing public and scientific opinion about the possibility of flying machines becoming practical. He died in 1896 of injuries sustained when his glider stalled and he was unable to regain control, having made more than 2,000 flights.

Lilienthal published useful experimental data on aerodynamics and inspired later pioneers. After Lilienthal, the development of a powered aeroplane was all but inevitable. He had demonstrated practical solutions to three of the four major design challenges: developing sufficient lift, achieving adequate stability and control, and building a light but strong enough structure. The fourth challenge was a lightweight engine, and the internal combustion engine was just becoming available around the turn of the century.

The Wright Brothers achieved the first controlled manned powered flight in 1903 and the first practical aeroplane in 1905. The Wright Brothers were bicycle engineers and adopted a

Fig 7.1: Lilienthal Glider. (Aviation Images)

methodical approach to development, building a series of gliders before making a machine with an engine. Being unable to procure a suitable engine from another manufacturer, they built their own, as well as the propellers to go with it, which was a major achievement.

Some of the Wright gliders were test-flown as unpiloted kites. They made use of Lilienthal's data and built a successful wind tunnel, which enabled them to measure the lift and drag of different wing shapes.

The invention of the aeroplane opened up the prospects of new applications with large markets. Within a few years of the Wright Flyer, just about every conceivable aeroplane configuration had been tried out, as shown for example in the 1965 British comedy film *Those Magnificent Men in their Flying Machines*. After just a few more years a particular configuration emerged as market leader. This was the fixed-wing biplane made of wood, wire, and fabric; and with a tail at the back and a propeller(s) in front of the engine(s) – the so-called tractor biplane. Many 'dead-end' developments were built and then discarded.

Fig 7.2: Wright Flyer. (Aviation Images)

The First World War provided a boost to development. Aeroplanes were used initially for reconnaissance and artillery spotting. Then their pilots began to try and shoot each other down, and specialised fighter aeroplanes evolved. Useful long-range bombers were just coming into service when the war ended. Airline flying on a significant scale started after the war using converted bombers. The London to Paris route was one of the first to attract a significant number of passengers.

The next major development, pioneered just after the war and beginning to mature in the mid-1930s, was the metal (predominantly aluminium alloy) monoplane with cantilever wings (i.e., no struts) and stressed skin (i.e., the skin of the structure bearing some of the loads.) There was little difference in weight between wood and metal structures, but metal was more repeatable in quality, less expensive to maintain, and was less expensive to manufacture if more than just a few production aeroplanes were required.

Variable geometry features such as cooling cowls in the engine mounting, variable pitch propellers, wing flaps, and retractable landing gear were individually pioneered earlier but had generally been dismissed as being too heavy and complex for the aeroplanes then being built, which cruised at around 100mph (161kph). However, higher powered engines enabled these features to be used together and aviation to enter a new phase, with cruising speeds about twice as high as before.

The Douglas DC-3, which first flew in 1935, had all of these features, and was the first commercial airliner capable of operating profitably without a government subsidy. It led to a rapid expansion of commercial flying, and was built in larger numbers than any other pre-war transport. It went on to become the most important Allied transport aeroplane in the Second World War. Even today, a few are in service.

The Second World War saw the introduction of the jet engine into early operational service, and the beginnings of swept wing development. The first aeroplane to exceed the speed of sound was the rocket-powered Bell X-1, which achieved this feat in 1947. Incremental improvements of these three technologies lead us to the present day.

The Boeing 747 first flew in 1969 and is still in production. It was the first airliner with operating costs low enough for mass travel.

Fig 7.3: Douglas DC-3. (Aviation Images)

Fig 7.4: Boeing 747. (Aviation Images)

Why was the aeroplane invented when it was?

Why was the aeroplane invented when it was, rather than sooner or later? One reason was the lack of a suitable powerplant before the invention of the internal combustion engine. A few pioneers tried lightweight steam engines, and John Stringfellow achieved the first powered flight with a model in 1848. But steam engines were too heavy for practical use.

This explains why the powered aeroplane had to wait for the internal combustion engine. But the materials needed for a glider (essentially bamboo, string and lightweight fabric) were available millennia earlier. So why did not the Romans for example build a glider? To answer this question, let us first consider what factors contributed to progress in the nineteenth century.

One factor was a shared vision of aeroplanes as a means of high-speed transport. Having conquered the oceans with the steel steamship and established high-speed land transport with the steam locomotive, the air was clearly the next place to try. Technical progress was 'in the air', so to speak. Towards the end of the century more than a dozen pioneers were trying to build the first aeroplane, mostly independently of each other.

Another factor was the emergence of engineering as an established profession, with the formation of learned bodies and an increasingly scientific approach. Technical journals helped to spread new ideas, thereby reducing the 'technology transfer gap'. The Aeronautical Society of Great Britain (later to become the Royal Aeronautical Society) for example was founded in 1866.

Octave Chanute, active in the second half of the century, played a key part in this communication process. He corresponded with most of the pioneers and disseminated the resulting ideas in a series of publications. From his civil engineering experience, he invented a braced biplane wing structure that became widely adopted.

Another possible factor is the development of the bicycle in the mid to late nineteenth century, although this is a speculative idea that has not been tested academically. Bicycles introduced welded tube structures and showed that unstable machines could be controlled. It may not be a coincidence that the Wrights were bicycle engineers, and that their early flying machines were unstable and difficult to fly. One of their pioneering contributions was the appreciation that pilots would need considerable training to fly an aeroplane.

These factors came together with the internal combustion engine, and the first powered aeroplane flew in 1903. If the Wright Brothers had not succeeded then, someone else all but certainly would have within a few years. Alberto Santos-Dumont flew his first aeroplane in 1906, just three years after the Wrights, and there were several others not far behind. Santos-Dumont worked independently from the Wrights, except that both corresponded with Octave Chanute.

None of these factors were available to the Romans. Even if they had built a simple hang glider, it would have been of little use to them. Engineers today take Newton's laws of motion for granted, and these are an essential part of any aeroplane design. But the Romans did not know these laws.

Airships are in effect streamlined balloons fitted with engines and control surfaces. They were just beginning to become useful when the aeroplane was invented and took over most of their potential tasks. The main problem with airships is their low speed, which makes them impractical to use in strong winds.

The Zeppelin raids on England in the First World War brought them into public prominence, but after a few were shot down the Germans switched to bomber aeroplanes. In the mid to late 1930s airships again came into prominence by providing the first regular transatlantic air service. However, flying boats were just a few years behind, with Pan American providing the first scheduled service in 1939; and some well-publicised disasters (such as the *Hindenberg* and the *R-101*) and the outbreak of war put an end to airship operations. The Second World War led to the construction of aerodromes all around the world and the development of reliable long-range landplanes, and these took over from flying boats as the primary means of providing long-range air travel.

What has been the influence of government?

During the pioneering stage, governments generally showed benign interest in aviation development. The potential advantages of the aeroplane for reconnaissance were fairly obvious to military and naval thinkers and, apart from the few inevitable blinkered naysayers, there was general support. In general, one can safely say that throughout aviation history, government backing has been 'about right', especially in terms of paying for national research facilities such as wind tunnels. Especially effective was the influence of the US Post Office Department in the 1920s in contracting air mail services to private airlines. This spurred the development of commercial US air travel.

But there have been cases where excessive political involvement has led to highly detrimental effects and two, one British and the other US, will be discussed here. *Empire of the Clouds* by James Hamilton-Paterson tells the dismal story of the failure of the British aircraft industry to live up to its post-Second World War promise. In the early post-war years, brilliant prototypes at spectacular air shows were heralded as showing Britain well in the lead, but were followed by lamentably few aircraft types built in significant numbers. Airliners of great technical promise had their sales severely curtailed by squabbling between industry, airlines, and government; or lost competitive edge owing to delays caused by management lacking the drive to make aeroplane developments happen on time. The Avro Tudor, Bristol Britannia, and the Handley-Page Hermes all failed to live up to their promise.

In 1952, the de Havilland Comet became the world's first jet airliner to enter service. However, a catastrophic design flaw led to several fatal accidents and grounding, and it was not until 1958 that it was ready to enter service again. Meanwhile, Boeing had taken the lead with their 707, which became the first successful jet airliner.

De Havilland's only previous metal airliner had been the pre-war Flamingo seventeen-seater, which first flew in 1938. This was a promising design but priorities of war led to only fourteen being built. Their only experience of jet-powered high-speed flight had been the Vampire fighter and the DH-108 research aeroplane. By contrast, Boeing had built several hundred large transports (the Stratocruiser and military equivalent), and large jet bombers (the B-47). Boeing was therefore far better prepared to bring in the jet age.

Of the all-new post war British airliners, only the Vickers Viscount sold in large numbers. This success was built on that of the preceding Vickers Viking, which was derived from the Wellington bomber of the Second World War. Several British 'workaday' commercial transports sold well, such as the Bristol Freighter and the de Havilland Dove.

The main cause of this disappointing achievement was the failure to reorganise the British aircraft industry after the Second World War. During the war, the industry had very close links with government. There was a cosy relationship in which developing a bad aeroplane did not cause severe hardship to the company that had built it. This arrangement worked well enough when designers, managers, and workers were highly motivated by the need to win the war, and when funding for better aeroplanes was all but unlimited. The UK did in fact produce several world-beating designs (the Spitfire, Mosquito, and Lancaster were arguably the best fighter, multi-role combat aeroplane, and heavy bomber of their day). But it took a long time for government and industry to learn the need for different disciplines in peacetime. Where performance had been paramount in wartime, the main post-war priorities became reliability, low operating cost, safety, and the availability of spares and technical support.

The aircraft industry ended the war with great prestige. The fighters had saved Britain from invasion and the bombers had spearheaded the attack on Germany. The companies were often led by the strong-minded pioneers who had founded them and who had no wish to be radically reorganised. Government had other priorities. So bureaucratic inertia reigned supreme and the industry carried on much as before with government sponsoring airliner development, and management often emasculated by cost-plus contracts. There was simply no need for them to become commercially efficient. It was not until the Duncan Sandys' defence white paper of 1957, which led to the cancellation of many projects and to large-scale reorganisation, that the situation began to change.

The underlying causes of excessive political interference go back two decades earlier. Britain ended the First World War with the world's largest air force and an industry making superb aeroplanes. The UK was the first to start an independent air force – the Royal Air Force founded in 1918. Other countries retained separate army and navy air corps. In this, the UK was indeed far-sighted. However, in order to maintain its position in the inter-service battles for funding between the wars, the RAF had to exaggerate the case for strategic bombing, which they claimed only they were equipped to do. This doctrine carried into the Second World War, in which vast resources were dedicated to designing, building, and operating heavy bombers. With not much hindsight, it would have been better to have built fewer bombers of higher quality and to have used them more selectively. The case for this was made at the time, but others had better access to Churchill's ear.

This overriding priority led to the UK giving up on airliner production and development during the war. Thus at the end of the war the UK had no up-to-date transports whereas the US had several well-proven and competent designs. So the post-war British aircraft industry had a double handicap – lack of proven modern transports and an industry too closely dependent on government.

Britain's aerospace industry has learned its lessons and is now second in size only to that of the US. Government now backs only projects that already have private sector backing. Prestige has given way to profit. It is less romantic but does provide many high-quality jobs and exports.

The second example of too much government involvement is the evolution of space launchers. Early satellites were launched using converted ballistic missiles, as these were the first artefacts capable of reaching space. It soon became apparent that throwing away a vehicle for each launch could never become economical, so, in the 1960s, most large aircraft companies had feasible plans to replace these expendable launchers with reusable ones like aeroplanes (spaceplanes). Had these been developed, there would now be an airline service to orbit and the cost of access to space would be about 1,000 times less than it is. But space policy was dominated by NASA, who have never given high priority to cost reduction, and orbital spaceplanes have not yet been built. But suppressed evolution leads to revolution, and spaceplanes are likely to transform spaceflight soon, as described in more detail in Chapter 11.

Government action has sometimes led to unintended consequences. For example, the Versailles Treaty of 1919 banned Germany from building powered aeroplanes. The Germans therefore took to gliding with great enthusiasm, which was a great help when it came to building up the Luftwaffe. To this day, Germany is arguably the world's top gliding nation. The Treaty did not ban rockets, probably due to an oversight. The Germans were thus the first to build a ballistic missile – the V-2 – which became the first man-made machine to fly to space and which became the progenitor of all ballistic missiles and launch vehicles since then.

Safety

A remarkable feature of modern airline flying is its safety. The accident rate for airline passenger flights in the late 1930s was roughly one fatal accident per 20,000 flights. Now it is one per several million flights. This improvement has been gained by painstaking incremental developments over the decades, such as more reliable engines and airframes, better weather forecasting, better navigation aids, radio landing aids, on-board radar for avoiding storms and collision with mountains and other aircraft, better air traffic control, better regulation, better understanding of how and why pilots and mechanics make mistakes, confidential incident reporting, and cockpit automation. Key elements have been international collaboration and the open dissemination of lessons learned from near misses and of course from accidents themselves. A major issue today is that cockpit automation is now so effective that pilots have little active flying to do and are at risk of losing the basic skills needed to cope with emergencies.

The influence of science

It is safe to say that the invention of the aeroplane provided a great stimulus to the science of aerodynamics rather than the other way round. The Wright Brothers, for example, knew from their wind tunnel that a wing with a high span produced less drag than one of the same area but with a shorter span. But they did not understand why, because their knowledge of aerodynamics was empirical only. They knew that the air forces acting on a wing were proportional to area and speed squared. They knew that lift increased more or less linearly with the angle between the wing and the airflow (the angle of attack) and reached a maximum at a certain angle (the stall angle), above which lift decreased and drag increased dramatically. They knew roughly the empirical constants in the resulting equations. But the underlying theories had not been discovered.

The explanation for a wing of high span having lower drag is given by the circulation theory of lift, published by F W Lanchester in several publications around the turn of the century, which was the first aerodynamic theory of practical use to designers. But when he tried to explain this theory to Orville Wright during the latter's visit to Europe in 1908, Orville could not understand what he was talking about! Lanchester was not very good at explanation, and Orville was very much a practical engineer.

This theory was slow to propagate, partly because of the First World War. The designers of the Sopwith Pup, a successful First World War fighter, for example, did not know about it. Ludwig Prandtl developed Lanchester's theory, and built up Göttigen as the leading centre of aerodynamic research. The first aeroplanes to benefit were late First World War German fighters.

Prizes

Prizes have played an important part in stimulating aeronautical development. The Schneider Trophy was announced in 1912 and was awarded annually to the winner of a race for seaplanes and flying boats. Landplanes had to operate out of small airfields, which meant that the wings had to be large enough to provide adequate lift at low speeds. Free from this limitation, seaplanes could be built with smaller wings, resulting in lower drag and higher speed. This trophy played a major part in stimulating the development of more powerful engines and more streamlined aeroplanes. If an aero club won three races in five years, they would retain the trophy. It was finally won in 1931 by a British Supermarine S.6B. Experience thus gained contributed to the success of the Supermarine Spitfire fighter, which played a decisive role in the Battle of Britain. The Schneider Trophy generated huge publicity for the time, with more than 200,000 spectators attending some of the races.

The Orteig Prize was offered in 1919 for the first pilot to fly non-stop from New York to Paris, or vice versa. Charles Lindberg, a pilot with the US Air Mail and US Army Air Corps reserves, won this prize in 1927. Lindberg became a ticker-tape hero, and the resulting publicity is credited with triggering a rapid growth of commercial aviation in the US.

The Ansari X Prize was announced in 1996 for the first private organisation to launch a reusable piloted spacecraft into space twice within two weeks, and was aimed to spur the developments of low-cost spaceflight. It was won in 2004 by the SpaceShipOne spaceplane, financed by Paul Allen and built by Burt Rutan's Scaled Composites company. Richard Branson's Virgin Galactic plans to use an enlarged development of SpaceShipOne to become the first to carry passengers on brief space experience flights. The first successful commercial operation of spaceplanes is likely to transform spaceflight by bringing in an aviation approach to replace the present throwaway launcher culture. This is discussed further in Chapter 11.

Further reading

Miller, Ronald and Sawyers, David: *The Technical Development of Modern Aviation*, (New York: Praeger 1970).

McCullough, David: *The Wright Brothers*, (Simon & Shuster, 2015).

Chapter 8
Modern Communications

Robin Morris

All communication between individuals before the nineteenth century depended, except for the written word, on direct visual or sound contact. The first record of a semaphore system was that made by R L Edgeworth in Britain in 1767 to relay the result of races at Newmarket. Then in 1790 C Chappe, a French engineer, set up a manually operated chain of semaphore stations spaced about ten miles apart which used telescopes to transmit coded flag signals between stations and was widely used in France during the Napoleonic wars, and continued in service until the 1850s. Such manual semaphore systems were costly in terms of labour, limited by adverse weather conditions, and confined to daylight hours. Significant progress, however, came only with the invention of electric telegraphy, which in turn depended on the invention by the Italian A Volta in 1799 of the electric battery, producing a constant current of electricity. Other highly important contributions to the development of early telegraphy include the invention of the moving coil galvanometer by J Schwigger in 1828, and the publication of the principles of electromagnetic induction by Michael Faraday in 1831. These steps launched modern electric communication.

Electric telegraphy

A stimulus to the invention of electric telegraphy was its potential use within the railway system, which began to grow rapidly in the 1830s. Its adoption in Britain was due to the collaboration of Charles Wheatstone and William Cooke, who had both been working independently on the problems involved. When visiting Heidelberg University in 1836, Cooke had witnessed Professor Munke demonstrating electrical signals passing by wire, and this had inspired him to construct a practical electric telegraph. Cooke formed a partnership with Wheatstone to produce this in 1837, and in 1838 they set up a thirteen mile telegraph connection between Paddington and West Drayton. It was then adopted officially by the GWR for its signalling system.

Another major contributor to the early development of the electric telegraph was the American S B Morse, the inventor of the Morse Code, which was first used with electric telegraph systems in 1844. His many inventions included a machine that recorded dots and dashes on a paper ribbon. A steady improvement in recording systems made it possible by 1860 to send and print up to sixty words a minute. By 1866 over 200,000 miles of telegraph line were in use in the US alone.

A highly important step was the invention in 1835 of the electromechanical relay by the American physicist Joseph Henry, which greatly improved the efficiency of the electric telegraph by strengthening the weak signals caused by line attenuation. The British scientist Lord Kelvin invented the mirror galvanometer about 1858, making it possible to detect very weak signals.

The invention of the electric telegraph led to a vast growth in the rapid transmission of information. Realising its potential, European nations were quick to install telegraph services, beginning with Austria and Prussia in 1849. Expansion of telegraph lines became extremely rapid, increasing in Europe between 1849 and 1864 from 2,000 to 80,000 miles (3,220km to 128,750km). In 1852 fewer than 350,000 calls were made, but by 1869 France and Germany alone had each sent six million calls.

The manufacture of cables to carry telegraph messages underwater presented a considerable technical challenge because of the need to maintain electrical insulation when exposed to very high water pressure. A start was made in India in 1838 when signals were successfully transmitted under the River Hoogly. In 1842 Henry More laid a wire insulated rubber cable across New York Harbour. Soon after, Charles Wheatstone laid rubber insulated cables in lead sheathing at a greater depth, but they leaked water. The Prince Consort was sufficiently interested in this experiment to correspond with Wheatstone, offering an interesting but impractical solution to the problem.

The difficulty was solved in 1847 by Werner von Siemens, who used a seamless coating of gutta percha as an insulator, and in the following year a two-mile undersea cable was laid in the North Sea by the Great Eastern Railway using this material. Being successful, this experiment opened up the prospect of laying lengthy underwater cables, so that the years 1850–70 saw a huge growth in underwater cable laying. In 1858, a cable was laid from Valencia in Ireland to Newfoundland, but it lasted only twenty days before the cable insulation failed due to the excessively high voltage. In 1866 a transatlantic cable was laid successfully by Brunel's SS *Great Eastern*, transmitting information at thirty-seven letters per minute. Telegraphic communication between Britain and India was achieved in 1870. Typical construction consisted of an inner copper core insulated with gutta percha, further protected by a covering of tarred hemp and wax encased in turn by an outer seamless metal sheathing.

The telephone

The idea of using sound waves to generate varying electric currents in order to transmit speech has been attributed to Charles Page in 1837, and in 1854 Charles Bourseul in France published a description of how speech might be transmitted electrically. Major problems were the lack of efficient microphones and speakers, and of connecting lines between subscribers. In the 1860s, the German J P Reis invented a diaphragm microphone, coining the German word 'Fernsprecher' and the English word 'telephone'. Meanwhile in America, from 1873 onwards, both Graham Bell and Elisha Gray were working on solving the problem of transmission of speech. Bell was granted an American patent for a transmitting and receiving system, a few hours before Elisha Gray. Bell also developed a system of sending more than one message simultaneously ('duplexing') and so saving the cost of copper cable.

The telephone soon began to challenge the telegraph service. For example, in 1880 there were 50,000 subscribers in the US, rising to 7 million by 1910. In 1896 telephone dialling was introduced in the US, replacing push buttons. In 1912, an automatic telephone exchange was developed in Sweden by G A Butulander and N G Palmgren. By 1914, telephone communication between the American east and west coasts had been established. The first teleprinter ('Telex') was operational in the US from 1928 and was used in Britain from 1932. By 1920, experiments had begun with telephone links using repeaters. The first transatlantic telephone cable ('TAT 1') was laid in 1956, running from Scotland to Newfoundland. It contained thirty-six telephone channels (two cables were used, one for each direction), using fifty-one repeaters. By 1976, the latest cable, TAT6, carried over a hundred times the capacity of TAT 1.

The American inventor Thomas Edison made an immense contribution to the development of telegraph and telephone systems, on which he spent much of his working life, taking out

1,180 patents. He developed a duplex system of telegraphy in 1873, and later a quadruplex system. His other inventions included a 'carbon button' transmitter in 1877 and a variable resistance microphone receiver in 1878. By then he had developed an automatic telegraph, operating at 500 words a minute, and amongst many other discoveries he claimed to have discovered thermionic emission.

Telephone communication systems

The earliest telephone communication system used was the 'crossbar', in which a switchboard operator made the connection manually by means of plugs to a horizontal bar from the vertical bars attached to the relevant lines. This system became more complex and labour intensive as the network grew in size. The first automatic switchboard was invented in 1899 by A B Strowger, an undertaker living in Kansas City. However, it was not widely used until 1921 in the US and 1929 in Britain. Importantly, this system introduced subscriber dialling. As distances between subscribers increased, problems arose due to signal attenuation. In 1894, an American Professor M Pupin invented the 'loading coil'. Inserted at intervals along the line, it prevented the distortion of signals caused by different frequencies travelling at different speeds. Another major invention was the telephone repeater ('signal amplifier'). The first thermionic repeater dates from 1915.

By the 1950s 'transistors' were being widely used in America for switching circuitry. The first commercial electronic telephone switching system was installed in 1963 by Bell Laboratories, although by 1977 only 6 per cent of lines in the US were switched electronically. The first electronic telephone exchange was installed in Britain in 1970 and the first computerised switchboard in 1976. In 1985, the first digital system using solid-state devices was introduced in the UK ('System X'). An important subsequent development originating in the US is the widespread use of telephone lines to interconnect computers using digital to analogue converters. This has enabled information networks to be built up, thus providing computer databases.

Radio

Between 1861 and 1862, James Clark Maxwell published a series of papers suggesting the possibility of generating electromagnetic waves. In 1887, Heinrich Herz discovered how to produce electrical oscillations and detect them at a distance, and the effect was greatly increased when the receiving circuit was adjusted so that its dimensions corresponded to the wavelength of the oscillations produced in the primary circuit. The first patent for wireless telegraphy was granted on 2 June 1896 to the Italian G Marconi. Events then moved rapidly and on 12 December 1901 Marconi achieved transatlantic contact, transmitting Morse signals from Poldhu in Cornwall to St John's, Newfoundland.

The thermionic diode rectifier was patented in Britain in 1904 by J A Fleming, and the application for a patent for a thermionic triode amplifier was submitted in 1906 by Lee de Forest in the US. Using his triode valve transmitter, he made a successful radio broadcast from the Eiffel Tower in 1908, the signal being detected 500 miles (800km) away. The First World War led to great improvements in thermionic valve technology, and on 14 November 1922 the BBC went on the air with 'LO' from London, quickly followed by Birmingham '5IT' and Manchester '2ZY'. Crystal detectors needing headphones were commonly used in the 1920s due to their low cost (about £5 upwards) compared with the early thermionic 'straight' two-valve receivers costing £25 upwards. Further advances in valve construction led to the Pentode valve by Tellegen and Holst in the Netherlands in 1928, enabling the manufacture of highly efficient superheterodyne receivers, which went into mass production in the 1930s. Car radios, already well established, then made their appearance in Britain. From 1946 onwards, miniaturised valves invented during the Second World War permitted the construction of portable 'handbag' radio sets and improved car radios. By 1956, portable transistor radios began to be made in Britain by Pye Ltd, but were then 50 per cent more expensive than contemporary valve radios.

Television

The first electric television, using Nipkow disc scanning, was built by A A Campbell-Swinton in Britain in 1908. The invention of the first storage camera tube ('iconoscope') by V K Zworykin in 1923 was an important step. Cathode ray systems used for television transmission were produced independently by Zworkin and P Farnsworth in the US in 1927. Cable television was also achieved in the same year by Bell Telephones, who in 1929 became the first to transmit colour television. In Britain, the BBC, after a period of appraisal, opened a television service in November 1936. It operated exclusively on 405 lines from February 1937 to September 1939, when it closed down because of the outbreak of the Second World War. Broadcasting resumed in 1945, remaining at 405 lines. BBC2 began a colour service in December 1967 using 625 lines, the first in Europe to do so. The change from analogue to digital television took place in Britain in 1970. Cable television did not experience any real growth in the US and Canada until the 1960s, but by 1982 there were 22 million subscribers in the US alone. The use of fibre optics made it possible to transmit TV and radio programmes and also telecommunications data through the same cables. The first television pictures were transmitted across the Atlantic in 1962 by TELSTAR 1, and by July 1963 sixteen European countries were exchanging television programmes with the US.

Radar

Heinrich Herz showed in 1887 that electromagnetic waves are reflected in a similar manner to light, but this phenomenon was not exploited until the 1930s. Britain was then in the forefront of radar development, the first effective system being developed in the summer of 1935. With the increasing prospect of war, technical advances were rapid. By 1939, America, Russia, Germany, France and Holland also possessed radar systems. By 1940, the whole of the east and south coasts of Britain were covered with a defensive chain against aircraft flying at 15,000 feet (4,572m) up to a range of 120–140 miles (193–225km). An important advance was the invention of the Cavity Magnetron by J T Randall and H A H Boot in 1940, which greatly improved signal output power and consequently better target definition. It was then possible to guide an aircraft using 'grid navigation' (GEE) to within two miles of its target up to a distance of 350 miles (563km). Important applications, both civilian and military, have been made over the following decades, increasingly in conjunction with digital computer systems. These have included air traffic control, aircraft identification, and military defence systems. Aircraft now fly with microprocessors controlling an increasingly wide range of internal functions.

Information Technology (IT)

Prior to the invention of the digital computer, information was only stored in analogue form. However, digital technology has led to accessing, recording and displaying information on a scale previously impossible. Consequently from the mid 1970s onwards, information has transformed the way society is organised. From banking, media and government down to the private citizen, all have become more organised and controlled through the medium of electronic communication. Perhaps because of this it is easy to forget that the driving force behind these changes has been the constant development of silicon integrated circuitry, itself the product of the earlier discrete transistor.

Following the invention of the germanium point contact transistor by Brittain and Barden at Bell Laboratories in 1947, transistors soon replaced thermionic valves within electronic equipment. Their advantages included being more robust and having a longer lifetime, less weight and smaller size. Steps in their evolution include mass production of germanium alloy transistors in 1952 at General Electric in America. Transistor radios followed in 1954, their number growing rapidly from then onwards. In 1954, silicon grown junction transistors, capable of operating at higher temperatures than germanium, entered production at Texas Instruments, Dallas.

An integrated circuit is one in which the functions of several solid-state components are integrated on a single semiconductor chip. The first germanium integrated circuit ('IC') was built at Texas Instruments by J Kilby in 1959. Meanwhile, an oxide masking process had been invented by C J Frosch in 1957, again in America. The ability to grow a stable oxide on silicon led to the invention of the silicon planar transistor and the silicon bipolar IC. From 1970 onwards, Metal Oxide Silicon ('MOS') ICs were also available which, although slower, were cheaper and more compact than bipolar devices. It was MOS technology that was soon to dominate bipolar in the field of large-scale integration ('LSI'). In 1962, S R Hofstein and F P Heiman at RCA built the first MOS IC. It used sixteen MOS transistors on a silicon chip. This was the catalyst that led to the development of micro-miniature solid-state systems. The first microprocessor was developed at Intel in 1971 by M E Hoff. It was a 4-bit IC processor (the 4004). The first programmable hand-held computer was built in 1972 by Hewlett-Packard.

Computers

The vast explosion of information technology has been largely due to the rapid development of computer systems, and began during the Second World War. The first computers were driven by vacuum diodes and valve amplifiers that dissipated considerable power. Thermionic devices, with their fragility and relatively short lifetime imposed practical limits, including size, weight, and circuit complexity. Subsequent development of computers was therefore restricted until the invention of solid-state transistors and diodes, which were smaller, more robust and used far less power per device. Since they greatly improved reliability at reduced cost, computing systems designed using discrete solid-state components were termed 'second generation'.

A further breakthrough took place following the invention of the silicon integrated circuit. Using planar technology led to even greater reliability at reduced cost whilst permitting greater complexity per unit area. Soon, the number of components per chip roughly doubled every year (this rate of multiplication became known as 'Moore's Law' after G E Moore, who coined it when employed by the Fairchild Corporation in America). Consequently, computing systems took up less weight and space whilst using less power per device and becoming ever more reliable, promoting a vast expansion in the use and versatility of what came to be regarded as the 'third generation' of electronic systems.

Optoelectronics

The modulation of light waves has found an important role in the field of long-distance communications. Success depended upon the development of low-loss fibres and reliable lasers. Two types of transmitter are used for optical fibre linkage: first, semiconductor Light Emitting Devices (LEDs), and secondly, Light Amplification by Stimulated Emission of Radiation (LASERs). Both types provide greater band-width than previous systems, and can carry much more information than radio signals. Optical fibre cables were developed in the early 1970s. They undergo much less attenuation than copper co-axial cables and require a much smaller cable diameter. The first fibre optic commercial systems were installed in 1977.

LEDs and lasers are now widely used in optical fibre systems. Lasers used in optical fibre optics cost more than LEDs, but are faster and more suitable for long range transmission. Laser action was first achieved using a ruby laser by H Maiman in America in 1960. Infrared radiation was observed from a forward biased Ga/As junction by J W Allen and P E Gibbons in the UK in 1960. Optical fibre communications were first developed by K C Kao and G A Hockham in 1966.

Communications satellites

The first communications satellite SCORE was placed in orbit by the US in December 1958. It transmitted taped messages for thirteen days. In January 1959 the Russian *Lunik 1* satellite was launched, followed by the American ECHO 1 in April 1960. By the mid-1960s television relays became routinely available using geo-stationary satellites. The first active repeater satellite was TELSTAR 1, which transmitted an image across the Atlantic in 1962. By 1984, over 500 satellites were in orbit. By 1966, direct inter-satellite communication had been achieved. By the late 1970s cable companies were offering a selection of channels beamed down from satellites. By then, high-powered satellites were transmitting TV signals strong enough to be picked up by dishes about one metre across. Sky began broadcasting in 1989. Interplanetary space communication began with the Russian MARS 1 satellite, which was launched in November 1962. Powered with solar cells, it lost communication with the Earth after 66 million miles (106 million km).

Global positioning systems (GPS) evolved steadily from the 1970s, and by March 1994 the US Air Force had established a fully operational twenty-four satellite GPS system. By this time there already existed an increasing number of commercial and military applications. The rapid growth of GPS systems may be judged from the fact that between 1996 and 2001 the number of automobile GPS devices in use increased from 1.1 million units to 11.3 million.

In 1998 the first habitable module (ZARA) of the International Space Station was set up. This formed part of a project involving collaboration between the US, Europe, Japan, Canada, and Russia. In 2000, a habitable control and command centre was included. By this time, the US, Europe, Canada, Russia, China and Japan all possessed satellites in orbit, recording weather, and for commercial and military applications. More recently, planetary satellites have made successful investigations involving the surface detail of our planetary system, including the furthermost – Pluto – in 2015. NASA's official plans in 2014 included sending a man to Mars, but the idea has met with increasing scepticism, due to cost.

The original World Wide Web ('WWW') prototype was written late in 1990 by T Berners-Lee, whilst working at CERN. He stated that 'the web's major goal was to be a shared information space through which people and machines could communicate'. He also expressed the opinion that 'This space was to be inclusive rather than exclusive.' What his achievement did was to enable any person or group with network access to share information with others, and may be viewed as a development of digital computer network technology dating from the 1960s. Subsequent development was extremely rapid and by the mid-1990s the WWW was already being used by millions. International electronic networks of this nature have had a highly significant effect on a global scale, including within the banking industry, and subsequently influencing profound economic consequences.

Constant shrinkage of semiconductor components (which have so far continued to obey 'Moore's Law') has so far (2016) permitted the construction of increasingly smaller communication systems whilst offering a far greater range of functions. Further component shrinkage using silicon planar technology is however now being challenged on grounds of physical limitations at quantum level (see IEE Spectrum, June 2015). Further advances in the field of nanotechnology (the design and development of materials and devices within the nanometre scale) may well extend to include the creation of increasingly complex biological systems.

The main obstacle to preventing even greater rapidity in shrinking mobile communications systems has been the failure to develop suitable power supplies. So far, Lithium-ion batteries continue to offer the best solution. They were introduced in 1991 and are now widely used in mobile phones, camcorders, and laptop computers. Current opinion (see Proc. IEEE, June 2014) is that it is too early to predict which cell chemistry may, if possible, offer the best alternative.

Further Reading

Day, L and McNeil, I: *Biographical Dictionary of the History of Technology*, (Routledge, 1996).

Antébi, E: *The Electronic Epoch*, (Van Nostrand Reinhold Co., 1982).

Randell, W L: *Messengers for Mankind*, (Hutchinson & Co.).

Ceruzzi, P E: *A History of Modern Computing*, (MIT Press, 2000).

Desmond, K: *A Timetable of Inventions and Discoveries*, (Constable, 1974).

Dummer, G W A: *Electric Inventions and Discoveries*, (IPP, 4th ed., 1997).

Braun, E, and Macdonald, S: *Revolution in Miniature*, (CUP, 2nd ed., 1982).

Morris, P R: *A History of the World Semiconductor Industry*, (Perigrinus, 1990).

CHAPTER 9

The History of Technology in Medicine

Richard Harvey

Technology did not have much impact on medicine until the nineteenth century. Before this, treatment was limited to the use of naturally occurring plants and minerals, and to simple surgical methods, based on observation and experimentation. This Chapter describes, in roughly chronological order, the increasingly technical ways in which medical treatments have progressed.

A medical tradition developed independently in many parts of the world thousands of years ago. In ancient Egypt and in Mesopotamia, India and China, the principles of taking a medical history, carrying out an examination, making a diagnosis and giving treatment and a prognosis were developed by scholars. The causes of disease were not known, and illnesses were ascribed mostly to supernatural influences, demons, magic or fate. Treatments included herbal medicine, application of creams and ointments, acupuncture (in China), massage and simple surgical procedures, such as removal of lumps, incision of abscesses and the treatment of wounds.

The causes of ill-health and death probably remained largely unchanged between the Neolithic period and the early part of the nineteenth century. Average life expectancy was low, probably in the region of twenty years at birth. However, many infants and children died within a few years of birth, and if these first few years of life were survived, life expectancy was much greater, perhaps forty-five years or more. The primary causes of death were infantile diarrhoea, smallpox, malaria, poliomyelitis, measles or tuberculosis, war, accidents, famine and malnutrition and death in childbirth. Such conditions are less common causes of death in developed countries at the present time, but still kill millions of people in many parts of the world.

The development of medical treatments

Rational medical treatment depends on an understanding of normal human physiology and the nature of disease processes.

However, many successful treatments were developed before such knowledge became available, being based on simple observation and experimentation. A few initial examples illustrate this. As knowledge of the fundamental mechanisms of disease processes increased, remarkable technological advances became possible, and some of these are described. This is a very large subject, so has been dealt with by giving selected examples of key medical advances.

Quinine

Malaria is an ancient disease, known in Asia, Africa and the Middle East and it was endemic in coastal areas of South East England from at least the fifteenth century, but was not differentiated from 'fever' of any cause. The bark of a South American native tree, cinchona, was used successfully by natives of Peru to treat fevers. Spanish Jesuit missionaries returning from South America introduced this treatment to Europe, and Robert Talbor, an English apothecary, cured many people living in the Fens and Essex marshes from fever, using a secret mixture based on cinchona bark, eventually treating Charles the Second and being appointed Royal Physician. After Talbor's death, the secret of cinchona and its active ingredient, quinine, became generally known. Quinine became a vital medication in the British colonies in Asia and Africa, greatly reducing the 50 per cent mortality previously seen in Europeans sent to the 'White Man's Grave' countries of West Africa, particularly Sierra Leone and the Gold Coast. Thus an effective treatment for malaria became available despite the fact that the cause of the condition and how the treatment worked were completely unknown. Two hundred and fifty years later the *plasmodium* parasite was identified and its life cycle and transmission by mosquitoes described, leading to alternative methods of attacking the disease, for example by targeting its insect vector, the female *anopheles* mosquito.

Smallpox

The story of the battle against smallpox also illustrates the value of simple observation, of experimentation and of application of successful methods to prevent a disease, despite little understanding of its cause. It includes the most dramatic triumph of medical science, the first complete eradication from the world of a killer disease.

For thousands of years, smallpox devastated mankind, even some Egyptian mummies showing the characteristic pockmarks.

Part of the decline of the Roman Empire has been attributed to the smallpox plague of Antonin, which killed more than 7 million people. Smallpox was introduced to the Americas by the Spanish and Portuguese, and had a devastating effect on the native inhabitants, being instrumental in the fall of the Aztec and Inca empires. In Europe in the eighteenth century, 400,000 people died of smallpox each year, the mortality rate being between one in two and one in five, with those who survived usually having disfiguring scars, and with blindness in about a third of cases.

It was well known that people who had recovered from smallpox were immune to further attacks, and could safely nurse those suffering from the disease. Because of this, the practice developed in India, China and Africa of giving people a (hopefully) minor attack of smallpox by inoculating their skin with a small amount of material from a smallpox pustule, a process called variolation. This technique was introduced to Europe at the beginning of the eighteenth century, probably by traders from Constantinople. About 2–3 per cent of people died of smallpox after variolation, but this mortality was less than a tenth of that seen with natural smallpox. Variolation became increasingly popular after being taken up by members of various European royal houses.

Edward Jenner started up in medical practice in Berkeley, Gloucestershire in 1773, after training with John Hunter in London. He had heard that milkmaids were immune to smallpox if they had previously caught cowpox, a mild infection prevalent in the cows they worked with. In 1796 Jenner met a young dairymaid with fresh cowpox lesions on her hands. Using matter from her lesions, he inoculated an eight-year-old boy, with no ill effects apart from mild fever. The boy was subsequently inoculated with smallpox, and appeared to be immune. Jenner called this process vaccination (from the Latin, vacca, a cow), and despite early controversy it spread rapidly throughout Europe.

In 1967, a global campaign of vaccination for smallpox was started by the World Health Organisation, and by ten years later the disease was eradicated from the world. In 1980, the World Health Assembly announced, 'The world and all its people have won freedom from smallpox', and recommended that vaccination should cease.

Scurvy

An unpleasant disease characterised by lethargy, skin rashes, bleeding gums, loss of teeth, breathlessness, bone pain, nerve damage and death, scurvy was common on long sea voyages, and was one of the limiting factors of sea travel, often killing large numbers of passengers and crew during long distance voyages. Between 1500 and 1800 it has been estimated that scurvy killed more than two million sailors. Scurvy was also described in the Crusades. Various fruits, acids and vegetables had been tried for scurvy, notably scurvy grass, but it was not until 1747 that James Lind, in the first ever clinical trial, showed that scurvy could be prevented and treated by supplementing the diet with citrus fruit, though not by other forms of acid. Scurvy was finally eradicated from the Royal Navy during the Napoleonic wars, when fresh lemons (and later limes) were issued to all sailors, with a remarkable health improvement.

A treatment for scurvy was thus developed without knowledge of the cause of the disease. Even in the twentieth century scurvy was still a problem, for example in various Antarctic expeditions, when despite knowledge of the eradication of scurvy in the navy by citrus fruit it was mistakenly thought that tainted tinned meat might be responsible. Members of both Scott's and Shackleton's expeditions consequently suffered from scurvy.

Although it was suspected that some chemical present in citrus fruits was responsible for preventing scurvy, it proved very hard to isolate it, because the sugars in the fruit interfered with the extraction processes. Finally, a Hungarian doctor, Albert Szent-Györgyi, who had been trying to identify the antiscorbutic chemical, and who happened to live in Szeged, the paprika capital of the world, tried paprika, which lacked the interfering sugars, and soon isolated and purified large amounts of 'ascorbic acid', enough to demonstrate that this was indeed the elusive substance, later called Vitamin C.

Digitalis

Until there was widespread use of penicillin in the 1950s, infections with *streptococcus pyogenes* were very common, typically as a sore throat. Many people developed a strange reaction to such infections, their body's immune response producing antibodies which attacked not only the bacteria but also various body tissues, particularly the joints, heart valves, skin and nervous system, giving rise variously to acute rheumatism, rheumatic fever, scarlet fever and chorea (involuntary twitching movements). A serious long-term consequence in many people was inflammation of the heart valves (rheumatic heart disease), in which the delicate valves became thickened and scarred and either narrowed or incompetent. This led to failure of the heart as a pump, with breathlessness, lethargy, increasing fluid accumulation in the body (oedema, dropsy, anasarca) and death. For centuries, heart failure was common and there was no effective treatment.

After medical training in Edinburgh, Dr William Withering began practice in Stafford, later moving to Birmingham General Hospital, where he built up a large practice, treating 2000–3000 poor people free in his clinics and travelling widely in the area. He published papers on botany and became a member of the Lunar Society of Birmingham, together with Joseph Priestly, Erasmus Darwin and James Watt.

In 1775, Withering was consulted by a patient with severe heart failure, but told him that there was no effective treatment available. When he met the man again by chance a little later, he was astonished to see that there had been a miraculous improvement, apparently due to a herbal remedy obtained from an old woman in nearby Shropshire. Withering found and bargained with her and got a list of the ingredients in her remedy, the active agent being leaves of the purple foxglove, *digitalis purpurea*. After extensive trials by Withering, which confirmed the effectiveness of digitalis in heart failure he introduced the treatment into clinical medicine in 1785. It is still widely used today, as the pure cardiac glycoside: digoxin.

Cholera

Cholera is an acute diarrhoeal disease, which rapidly leads to massive loss of fluid with dehydration, circulatory collapse and death in up to 10 per cent of cases. The disease has probably existed since ancient times, with a main focus in the Ganges region of India. In 1817 millions died in a pandemic of cholera affecting India, China, Japan, Southeast Asia and the Middle East. A second pandemic started in 1826 in Russia, spreading to the rest of Europe, North Africa and North America, and reaching London in 1832. Initial cases were usually in seaside towns, reflecting the arrival of infected people by sea. The cause of cholera was thought to be bad air, 'miasma', in the atmosphere, although a Dr John Snow suggested that contaminated water might be responsible (an idea that received no support). In 1854 another epidemic occurred in London, with over 10,000 deaths. There was a severe outbreak in Soho, where John Snow lived in Frith Street, and in the next three days 127 people living in the Broad Street area died and more than 500 died within three weeks. John Snow interviewed all of the affected families in the area and marked their locations on a map. The central point of the epidemic was shown to be a pump at the corner of Broad Street and Cambridge Street. He persuaded the parish authorities to remove the handle from the Broad Street pump and within two weeks the outbreak was over. The puzzling deaths of a woman in Hampstead and her niece were explained by the woman's son, interviewed by John Snow, who said that she had water from Broad Street sent up specially, because she liked the taste. The contamination was probably due to leakage into the well of fluid from a contaminated cesspool nearby.

Further progress in the understanding and management of cholera depended on the isolation of the causative organism *vibrio cholerae* by Robert Koch in 1883, the demonstration that mortality can be greatly reduced by giving plentiful fluids to sufferers (oral rehydration therapy) and the recognition that the provision of clean water supplies can prevent the disease. Despite this knowledge, the World Health Organisation estimates that at least 750 million people in the world do not have access to clean drinking water, and each year there are between 100,000 and 600,000 reported cases of cholera globally, and probably many more unreported cases. A frustrating failure.

Puerperal fever

Childbirth was for centuries a hazardous event, and many women died of childbed (puerperal) fever. In 1844, Ignaz Semmelweis became an assistant in the Vienna General Hospital, specialising in midwifery. There were two maternity clinics in the hospital, one staffed by midwives and the other by medical students. The death rate in the first clinic was 2 per cent, whereas that in the second clinic was 13 per cent. There was no obvious explanation for the difference.

A friend of Semmelweis cut himself while dissecting a cadaver, the wound became infected and he died soon after, with symptoms very similar to those of puerperal fever. This observation led Semmelweis to believe that some infectious substance was being carried by the medical students from cadavers to the labouring mothers. He therefore insisted that the students should wash their hands in chlorinated lime solution before attending the maternity unit. Deaths from puerperal fever in this unit immediately fell from 13 per cent to 2 per cent.

These findings were not well received by his colleagues or other medical professionals, partly because they did not like the idea that they might be responsible for some of the fatalities. Although his findings were eventually accepted, before this the continuing criticism made him increasingly depressed, and he died in an asylum at the age of forty-seven.

Germs and magic bullets

Louis Pasteur was appointed Professor of Chemistry at the University of Lille in 1854, and was asked to find solutions to the practical problems of the local wine industry. He was able to show that bacteria were responsible for souring wine (and later similarly for souring milk). He showed that this process could be stopped by heating and then subsequently cooling ('Pasteurisation'). He also showed that the bacteria responsible were introduced from the air, which was disputed by others who believed that they were generated spontaneously. Pasteur promoted the germ theory of disease, which proposed that germs attacked the body from outside, a theory that initially was widely disbelieved. However, Pasteur developed this theory to explain the causes of many diseases, particularly anthrax, cholera, tuberculosis and smallpox.

In 1876, Robert Koch in Germany isolated the anthrax bacillus (*bacillus anthracis*) from an infected cow, was able to grow it in pure form in culture, and then by inoculating the culture into animals could reproduce the disease, a remarkable achievement according to Pasteur. However, tensions following the Franco–Prussian war of 1870–1871 meant that Pasteur and Koch were to be rivals rather than collaborators.

Koch next turned his attention to tuberculosis, and in 1882 reported that the disease was caused by an infection with a bacillus, *mycobacterium tuberculosis*. A lecture given by Koch in Berlin on this subject was attended by Paul Ehrlich, a doctor working at the Charité, Berlin, who had been using dyes to stain tissues and blood cells. He improved Koch's method for staining *mycobacterium tuberculosis*, and the two became friends. Ehrlich introduced in-vivo staining, in which dyes were injected into living animals, and their distribution studied. One of the dyes that he used was methylene blue, and since the malaria parasite could be stained with methylene blue, he thought that there might be an antimalarial effect, which there was (although treatment of humans with methylene blue would have been impractical). He experimented with many dyes, and found that guinea pig trypanosomiasis could be cured by trypan red. An aniline dye combined with arsenic was tried as a cure for human trypanosomiasis (sleeping sickness), but although effective it was too toxic. Ehrlich and his chief biochemist experimented with large numbers of similar compounds, and the 606th of these, arsphenamine, was found to be ineffective against trypanosomes, but very effective against *treponema pallidum,* the causative organism of syphilis. Marketed as Salvarsan, this was the first 'magic bullet', by virtue of its specificity against syphilis.

In a search for other 'magic bullet' compounds, other dyes were investigated. Hundreds of coal tar-derived azo dyes (containing the chemical azo group) were investigated by the German company Bayer AG. After years of fruitless trial-and-error work, a red dye that appeared active against streptococcal infections in mice was discovered. Marketed in 1935 as Prontosil, this was the first of the sulphonamides. Its active metabolite, sulphanilamide, and other 'sulfa' drugs were the treatments of choice for bacterial infections until the advent of penicillin, and are still in widespread use today.

Penicillin

Dr Alexander Fleming was working as a bacteriologist at St Mary's Hospital, London, when in September 1928 he noticed that mould had grown on some of his bacterial culture plates while he had been away on holiday. There was a clear space around one of the mould growths. Fleming realised that the mould, *penicillium notatum,* was producing a chemical which inhibited bacterial growth, and which could be found in a filtered broth culture of the mould. He coined the term penicillin for the active substance, and the word antibiotic to describe its antibacterial action. Fleming was not a very good communicator, and did not test the effects of penicillin *in vivo,* for example on infections in mice, so there was little interest in his findings.

With the outbreak of the Second World War, a team of researchers at Oxford led by Howard Florey, an Australian, and Ernst Chain, a refugee from the Nazis, demonstrated the effectiveness of penicillin on infections in animals and humans. Methods of large-scale production were developed, so successfully that over two million doses were available in time for the Normandy landings in 1944. Penicillin became the first of a series of antibiotics produced by one organism that were effective at killing other organisms. Streptomycin, active against *mycobacterium tuberculosis*, was the next major antibiotic to be introduced into clinical medicine, in 1954.

Targeted drugs

By the 1960s, advances in knowledge of human physiology led to a new approach to drug discovery. Most processes in the body are controlled by chemicals that interact with cells by attaching to specific receptor sites on the cell surface. In the brain and nervous system there are a large number of 'neurotransmitters', which act at nerve junctions in a similar way.

Adrenaline is a simple molecule with widespread effects. It is released at nerve endings of the sympathetic nervous system and also is secreted into the bloodstream by the adrenal gland and carried to its target organ receptors by the blood stream. It is responsible for 'fight or flight' responses, for example increasing heart rate and blood pressure.

Dr James Black, working at Imperial Chemical Industries, wanted to develop a drug that would be helpful in the treatment of angina by decreasing the work of the heart and therefore its demand for oxygen. Heart muscle had been shown to carry two different receptors for adrenaline, α and β receptors, so blocking either might be effective treatment. In due course a new compound, propranolol, was synthesised by Black and his team, and found to be a potent blocker of the α receptor. It was a very effective treatment for angina, (and revolutionised treatment of this condition), and for high blood pressure and disorders of heart rhythm, and rapidly became the world's best-selling drug. Other α and β blockers followed, with differing affinities for α and β receptors, notably the very successful atenolol.

Not content with this success, Black wanted next to try to find a similar blocking agent for the histamine receptor responsible for stimulating acid secretion by the stomach. Imperial Chemical Industries were not interested in this project, so Black left to join a rival company, Smith Kline and French. Over the next twelve years, Black developed his second revolutionary drug, the histamine H_2 receptor blocker cimetidine. This was found to markedly inhibit gastric acid secretion, with dramatically beneficial clinical results in patients with duodenal or gastric ulcers or with oesophagitis due to acid reflux. It was hailed as a wonder drug, and within a short time overtook propranolol as the world's best selling prescription medication.

Molecular biology and gene therapy

Many diseases result from abnormalities in the genetic DNA within cells. These may result in a problem with the coding for production of a specific protein, leading to a deficiency of that protein, for example the bleeding disorder haemophilia A, which is due to failure to produce the clotting factor Factor VIII. There may be abnormalities of genes that control cell growth and replication, as in cancers and leukaemias. The possibility of modifying human chromosomes to treat disease has been explored in recent years. There are two main possibilities, adding a gene to replace one that is not working, or else disrupting the action of genes that are not working properly.

Advances in molecular biology mean that many genes involved in disease processes have been identified and their DNA sequence can then be synthesised to form an artificial gene. In order to have an effect, the synthetic gene has to be introduced into the inside of cells, which is generally achieved by incorporating it into an engineered virus, which serves as a vector, carrying the gene through the bloodstream, across the cell wall and incorporating it into a chromosome. These techniques are too complex to describe in detail here. Many clinical trials are under way, particularly concentrating on conditions where there is a single gene defect, such as haemophilia, thalassaemia, immunodeficiencies and cystic fibrosis. The therapeutic gene is transferred to the somatic (non-sex) cells of the recipient, so can only treat that individual and cannot be transmitted to future generations. In contrast to this somatic gene therapy, germline gene therapy involves modifying germ cells, either eggs or sperm, so that all of the cells in the body that are derived from these cells will carry the modified gene, which will be passed on to future generations, in theory eliminating a hereditary disease. Ethical considerations have led some countries to ban such treatment, at least for the moment, but the prospects are encouraging.

The development of surgery

Before the nineteenth century, two main problems limited anything other than the most basic and simple surgical treatment, firstly the pain of a surgical procedure and secondly the high risk of wound infection, with often fatal results.

Pain relief during surgery

For centuries, opium or (after the seventeenth century) an opium-containing mixture, laudanum, was used, and the suffering of the patient was mitigated by the speed of the surgeon, for example in amputating a diseased limb in under a minute. In 1798, Humphrey Davy, at the Pneumatic Institute in Hotwells, Bristol, experimented with nitrous oxide ('laughing gas'), and noted that it had analgesic properties that he thought might be useful. It is still used in dental practice and as 'gas and air' in obstetrics, though the effect is not strong enough for major surgery. Boston dentist William Morton publicly demonstrated ether as a surgical anaesthetic agent in 1842, and chloroform was introduced by James Simpson in 1847. Both agents relieved pain but also rendered the subject unconscious, a big advantage during a frightening procedure. In 1853, chloroform was administered to Queen Victoria during the birth of Prince Leopold, and it quickly came into widespread use, enabling longer and more complex operations to be performed. Nowadays newer and safer halogenated hydrocarbons are the key ingredients of modern anaesthesia.

Combating wound infection

Joseph Lister, a Scottish surgeon, read about Louis Pasteur's work showing that rotting and fermentation of foods were due to microorganisms, and that these could be destroyed by filtration, heat or chemical solutions. He thought that something similar might be happening in infected wounds. Lister tested the results of spraying surgical instruments, surgical incisions and dressings with a solution of carbolic acid and found a remarkable reduction in the rate of gangrene. He published in 1867 the results of a series of studies that confirmed the effect of such 'antiseptic' treatment. Subsequently it was reasoned that rather than killing bacteria which had entered a wound, it would be better still to prevent them entering the wound in the first place. This led to the development of 'asepsis', which is a basic tenet of modern surgical practice.

Complex surgical procedures

Advances in anaesthesia, aseptic technique and the development of blood transfusion and fluid replacement therapy have made longer and more complex surgical procedures possible. In the course of the twentieth century, surgeons became specialised in particular areas of surgery, such as general (mainly abdominal) surgery, orthopaedics, cardiothoracic surgery, neurosurgery, plastic surgery and so on. The Royal College of Surgeons of England lists ten well-defined surgical specialties for which specific higher training is required.

Each surgical specialty has developed highly technological methods, which are too numerous to be dealt with in detail here.

The general surgeons developed 'minimal access' ('keyhole') surgery, particularly for abdominal operations, so that, for example, removal of a diseased gallbladder for gallstones (cholecystectomy) used to be done through a six inch incision below the ribs and need a two week hospital stay afterwards, whereas with keyhole surgery two or three tiny incisions are made and the patient goes home on the same or the next day.

Orthopaedic surgeon John Charnley, at Wrightington Hospital in the 1960s, pioneered the replacement of a damaged hip joint (a ball-and-socket joint) with an implanted artificial hip consisting of a stainless steel femoral stem with a round head and a polyethylene socket attached to the pelvis with bone cement. Various improvements have been made since that time, but the basic design remains the same. Other joints also can now be replaced, notably the knee, shoulder and ankle. Hip replacement is now the commonest orthopaedic operation, with about 100,000 procedures a year in England and Wales. The numbers of joint replacements is steadily increasing, reflecting the increasing numbers of elderly people.

Cardiac surgery generally depends on the surgeon having access to a non-beating heart, which requires cardiopulmonary bypass. A heart-lung machine artificially circulates and oxygenates the blood during the operation. These complex techniques were developed during the 1950s and 1960s. Blood is withdrawn from the patient's circulation via a large cannula placed usually in the vena cava or femoral vein and led via a pump to a membrane oxygenator, which replaces the gas exchange function of the lungs. A microporous hollow fibre gas-permeable membrane separates the blood from the oxygen, allowing gas exchange to take place in a relatively non-traumatic fashion. Oxygenated blood is then returned to the arterial circulation of the patient via a second cannula, usually into the aorta. Restarting the heart and cardioversion with a DC shock to correct abnormal rhythms such as ventricular fibrillation if necessary are routine. Surgical procedures in which cardio-pulmonary bypass is used include coronary artery bypass surgery, heart valve repair or replacement, closure of atrial or ventricular septal defects, surgery to congenital heart defects and heart or lung transplantation. Transarterial procedures using steerable catheters are generally carried out by cardiologists or radiologists rather than cardiac surgeons.

Neurosurgery is particularly problematic, partly because tissues in the brain and spinal cord do not generally heal well or regenerate, and partly because many vital areas of the brain are so close together that operating on one area risks damaging an adjacent one. This means that malignant brain tumours (gliomas) are generally incurable, although radiotherapy and chemotherapy may give some palliation. Neurosurgery within the cranial cavity is most effective when not operating for malignant disease, particularly for the removal of pituitary adenomas, clipping of intracranial aneurysms, placing of shunts in the lateral ventricles for hydrocephalus and stereotactic surgery for Parkinson's disease. Spinal canal surgery for spinal cord or nerve root compression is safe and effective.

Plastic surgical methods were used in ancient times in Egypt and India, for example to change the shape or size of a nose. Only minor procedures could be done until the advent of asepsis and modern anaesthesia. Advances in plastic surgery were stimulated by the facial injuries suffered by soldiers in the First World War (Sir Harold Gillies), and then by the burns suffered by aircrew in the Second World War (Sir Archibald McIndoe). There are many subdivisions of plastic surgery, now including the rapidly growing area of cosmetic surgery, which is carried out on normal parts of the body with the sole aim of improving a person's appearance.

Transplants and implants

Many body organs can fail through disease or injury. Although their function can sometimes be reproduced by machines (for example, dialysis in the case of the kidney), the function of many organs such as the liver, heart or lung is so complex that machines cannot reproduce them for anything but the shortest of times. Now that the problems of immunological graft rejection have been overcome by tissue matching and immunosuppression, transplantation of such organs from brain-dead donors offers the chance of life to patients who would otherwise die. Living donors can also donate some organs, for example bone marrow, a kidney or a portion of their liver. These techniques are beset by ethical and technical problems, and in the long term hope lies in the possibility that stem cell methods may succeed in growing the relevant organ in vitro.

A wide variety of manufactured implants is now available, including artificial joints, heart valves, pacemakers, arterial stents, aortoiliac grafts, grommets, cochlear implants, gastric bands, cardiac septal defect closure devices, ventricular shunts and lenses.

Imaging

The diagnosis and management of disease depends on accurate knowledge of internal anatomy. Before the twentieth century this was not possible. However, over the last hundred years some extraordinary technological advances, many using complex and very expensive technology, have revealed in minute detail much of the internal structure and many functions of the body.

In 1895, Wilhelm Roentgen described his experiments with a Crookes (cathode ray) tube and a fluorescent screen coated with barium platinocyanide. He reported that some unknown invisible rays (he called them 'X' rays) were being produced in the tube and could pass through paper, cardboard and even books to cause illumination of the screen. He found that he could obtain an image of his wife's hand on a photographic plate using these X-rays.

Soon after this, Henri Becquerel discovered that uranium produced rays similar to X-rays in their ability to penetrate various materials, and Marie Curie decided to carry out research into this phenomenon. Analysing the uranium ore pitchblende, she discovered that elements other than uranium also possessed what she called 'radioactivity', and she then isolated thorium, and later polonium and radium. She realised that the production of these rays was spontaneous and must be coming from the atoms themselves.

Both X-rays and radioactive isotopes have established major roles in modern clinical practice. In 1914, at the onset of the First World War, Marie Curie became Director of the French Red Cross Radiology Service and set up twenty mobile X-ray units, which immediately proved their value. Further modifications of technique allowed visualisation of the digestive tract (after instillation of radio-opaque suspensions such as bismuth and later barium) and led to X-ray tomography (a method of focussing X-rays) and latterly computerised tomography (CT).

Isotopes (particularly ^{99m}Tc, which emits Y rays) are used in scanning organs such as brain, bone, thyroid and liver, after being attached to material taken up preferentially by these organs. Dynamic studies can be done as well, for example isotope renograms for kidney function or thallium for scanning of the heart.

Radioactive isotopes can be used to measure very small amounts of substances in blood, using either saturation analysis (for example vitamin B_{12}, using B_{12} labelled with radioactive ^{57}Cobalt and a specific B_{12}-binding protein) or radioimmunoassay (for example insulin, using insulin labelled with ^{125}Iodine and specific anti-insulin antibodies).

Early workers with X-rays and radioactive materials found that skin rashes like burns (sometimes with ulceration) or hair loss could develop. X-rays were then used successfully to treat skin cancers, and radium was also shown to be effective. After a period of great excitement during which such 'radiotherapy' was used for all sorts of conditions, including tuberculosis, mostly with little success, and radium was sold as a panacea, it was realised that irradiation had some serious side-effects. In the 1920s there were deaths among girls painting watch and clock dials with luminous paint containing radium. Cases of carcinoma developing in patients treated with irradiation were reported, often after a delay of many years.

Nowadays radiation therapy is a valuable adjunct to the treatment of many cancers. External beam therapy is most often used, with radiation from an external source being directed at the target, usually in a series of treatment courses. Other types of external beam therapy are being developed, for example using proton beams rather than X-rays. In contrast, $Yttrium^{90}$, a source of β-radiation (electrons) can be used internally, either as yttrium needles placed directly within a tumour, or else bound to specific monoclonal antibodies which attach themselves to the target tumour. Because the electrons have a very short range in tissue, the effect is very local.

The hazards of irradiation are now well known, so diagnostic X-ray dosage is kept as low as possible, particularly in vulnerable groups such as children and pregnant women.

Positron emission tomography (PET) scanning is a very sophisticated functional imaging technique, in which a radionucleide in trace amounts is linked to a biologically active molecule, and its distribution in the body is mapped out. A common trace substance is fluorodeoxyglucose, which is taken up in areas of increased metabolism and glucose uptake. This is used clinically to detect cancerous metastases, but is also a valuable research tool.

Ultrasound consists of sound waves at a frequency too high for the human ear to hear. Ultrasound waves bounce back from surfaces and interfaces, and can then be recorded in picture form. Generated in a piezoelectric transducer, repeated short pulses of ultrasound are applied to the surface of the body and internal structures can be seen in great detail. Since no ionising radiation is involved, the procedure is safe, particularly in pregnancy. Obstetric ultrasound was popularised by Ian Donald, Regius Professor of Midwifery in

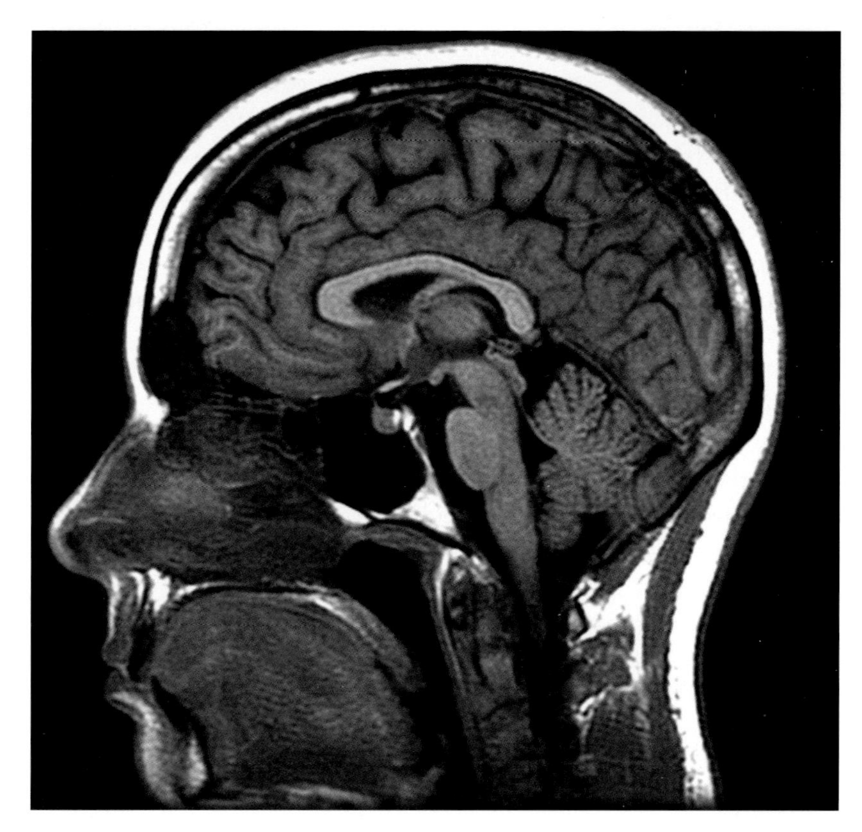

Fig 9.1: Magnetic Resonance scan of the human head. (*Richard Harvey*)

Glasgow, in the 1950s and 1960s and has become indispensible in the management of pregnancy. In addition, dynamic studies can be carried out using Doppler measurements, which can measure flow in blood vessels and through the heart ('echocardiography'). Small blood vessels can also be visualised, after injection of microbubble-based contrast media, and newer techniques continue to be developed.

Magnetic resonance imaging (MRI) scans give the most detailed of any scans of internal organs. Strong magnetic pulses polarise the natural 'spin' of various atoms, particularly hydrogen, which as water is present in large amounts in the body, and after the pulse, as the spun atoms relax back to their previous state, a faint radio wave is produced and can be detected. Complex computer analysis then can produce remarkable pictures.

Endoscopy and interventional radiology

Until the 1960s, direct visualisation of the gastrointestinal tract was limited to the oesophagus, a small area of the proximal stomach and the rectum. Rigid metal tubes could be introduced (the first time the stomach was reached was in a professional sword swallower), but could not see round corners. The Hermon–Taylor gastroscope had a flexible end with lenses, allowing visualisation of the distal stomach, and a gastrocamera with a flexible tube and a lens at its end with a length of miniature film, exposed in a series of pictures by light flashes, were used for some time. Fibreoptic technology was the breakthrough that allowed a large family of endoscopes to be developed through the 1970s – gastroscopes, duodenoscopes (side viewing, allowing cannulation of biliary and pancreatic ducts), long enteroscopes for the small intestine and colonoscopes. All of these consisted of a coherent bundle of glass fibres which carried the image, a separate smaller bundle which carried illumination from an external light source, a channel for suction or inflation and another for passage of instruments – biopsy forceps, snares for the removal of polyps, cutting wires and baskets for removal of gallstones from the common bile duct, inflatable balloons (for the dilatation of strictures) and lasers. Fibreoptic endoscopes have now largely been replaced by video endoscopes, basically of the same design but with the image derived from a small video camera on the end of the endoscope and displayed on a television screen. Similar instruments are available for other regions of the body, for example the nose, respiratory tract and bladder, and for paediatric use. Others can be used through small incisions, arthroscopy, laparoscopy or thoracoscopy, for example.

Access to internal organs and spaces using small steerable instruments has blurred the distinction between specialties. Thus, operative procedures using gastrointestinal endoscopy can be performed by physicians, surgeons or radiologists, given appropriate training. Some procedures, for example ERCP (endoscopic retrograde cholangiopancreatography), in which a contrast medium is injected via a side-viewing duodenoscope into the bile and pancreatic ducts, require X-ray screening. This means that it is convenient for a radiologist to carry out the whole procedure. Other minimally invasive interventions needing screening are also carried out by radiologists, for example the placing of a drain into an abdominal abscess, intravascular treatments such as coronary angiography, angioplasty, placing of stents, embolisation of intracranial aneurysms and TIPSS (transjugular intrahepatic portosystemic shunt) to reduce portal blood pressure in patients with bleeding varices. Those manoeuvres that are related to the heart are also carried out by cardiologists, particularly coronary artery stenting and ablation of accessory conducting bundles to treat various abnormalities of heart rhythm.

Summary

After steady progress in the application of technology to medicine, there has in the last twenty years been a remarkable acceleration. Hugely complex machines and treatments are now available, but the main limitation to their use is the cost (see Table).

Table: 2018 Costs of some advanced medical technology

Equipment	Function	Approximate cost (£)
Cobalt machine	Radiotherapy	5,000
Heart-Lung machine	Cardiopulmonary bypass	15,000
Image intensifier	X-ray screening	30,000
Linear accelerator	Radiotherapy	35,000
Gamma camera	Isotope scanning	40,000
Endoscopy equipment	Endoscopy	50,000
CT scanner	Imaging	100,000
DNA sequencer	Molecular medicine	150,000
MRI scanner	Imaging	750,000
PET tomography	Functional imaging	2,500,000
Proton beam generator	Radiotherapy	200,000,000

Unfortunately, the pace of advances in medical technology is greater than our ability to pay for them. Paradoxically, as developed countries have seen the virtual disappearance of many diseases, and molecular biology holds the promise of cure of many others, new largely self-inflicted problems have moved to centre stage. Sedentary lifestyles, smoking, excessive alcohol consumption, drug use and obesity seem set to afflict future generations.

Further reading

Brodsky, I: *The History and Future of Medical Technology*, (Telescope Books, 2010)

Kramm R, Hoffmann K P, Pozos R, (eds): *Springer Handbook of Medical Technology*, (Springer Verlag, Berlin, 2011)

Rubin R P: 'A Brief History of Great Discoveries in Pharmacology', *Pharmacological Reviews* 2007: 59; 289–359

Gawande A: 'Two Hundred Years of Surgery', *New Eng J Med* 2012:366;1717–1723

Technology and Medicine. www.sciencemuseum.org.uk

Chapter 10

Technology and Society

Angus Buchanan

Fig 10.1: Segovia Aqueduct. The Roman aqueduct at Segovia in Spain is one of the many surviving remains of the elaborate water-supply systems constructed in the Roman Empire. (*Angus Buchanan*)

When the nuclear reactor at the Chernobyl power plant in the Ukraine exploded in April 1986, it became abundantly evident that the production of electricity from nuclear fission was an industry based on a hideously dangerous process. The lesson, such as it was, was reinforced when a more advanced reactor at Fukushima in Japan was overwhelmed by a tidal wave following an earthquake and tsunami which partially flooded the plant in March 2011, causing another release of highly toxic waste that rendered a large area uninhabitable and released a nuclear plume into the atmosphere. With every possible attention to safety precautions, it is clear that a significant element of danger remains in this industry, which could result in an international disaster of unimaginable extent. Yet the world has become dependent on this source of power to make up the growing deficiency in alternative methods of generating electricity. This, in a nutshell, is the nature of the most serious physical problem confronting the modern world, and it demonstrates beyond any possibility of denial the intimacy of the relationship between technology and society. Technology is fundamentally a social concept encompassing the means by which people respond to their needs and resources, so that it is vitally important to have a clear view of the relationship between technology and the rest of society.

The idea of society

The word *society* has been used in a wide variety of senses to describe forms of human associations, from very loose collections of individuals, such as people attending a football match, to highly organised and complex bodies such as churches and states. Some commentators have dismissed the term as so wide as to be useless; but this is unreasonable because human beings are by nature sociable, needing to establish relationships with other individuals for survival and the enjoyment of life. All such associations have a degree of collective purpose, and some agreement about ways of achieving that purpose – even amongst members of a football crowd – can therefore be regarded as a 'society'. Some, to be sure, are obligatory societies, in the sense that we are born into them or find them imposed on us; but most are voluntary as we can choose them for instruction, sport, or creativity, and then leave them when we wish to do so. By far the most important of the obligatory societies – apart from the family, into which we are born – is the state, which is a discrete society with the distinctive function of being responsible for preserving the integrity of the group against both attacks from outside and disintegration from inside: in other words, for defence and for the maintenance of law and order. Everybody is born into such a society and has an obligation to abide by its rules or laws, whether it is a small tribal group or a large national association. It is possible to opt out of membership of some states, but usually only if another state is joined: the alternative, of being stateless, has long been recognised as a condition of life being 'nasty, brutish, and short'.

Technology has always played a vital part in the formation and preservation of states, and with the rise of the modern nation states this role has become crucial – and problematic. We have harnessed powers unimaginable to our predecessors until the very recent past, and with them we have created a global civilisation of enormous wealth within which most people are able to live with a decent diet, to travel with amazing facility to any part of the world in a short space of time, and to communicate instantaneously with friends and acquaintances over any distance. At the same time, we have used our technological powers to promote intense rivalries between different members of the world community, only narrowly avoiding the ultimate horror of atomic warfare, and in the process we have put at risk the survival of civilised life by the pollution of our environment, the depletion of irreplaceable resources, and the almost unbearable pressure of population increase. Similarly, we have failed to ensure a fair and even distribution of the fruits of our technological successes, so that problems of mass poverty and ignorance persist within the brilliant cornucopia of technological benefits. It seems that we have triggered changes in the atmosphere of our planet which are causing unusually rapid climate changes and rises in the sea level that are a threat to the survival of many of our great cities. All this, in brief, is the challenge with which our technological competence presents us in the twenty-first century. Can these tensions be resolved in time to prevent the self-destruction of our advanced technological civilisation?

Answering this question requires some deconstruction. It does not imply that technology itself is responsible for the problem. The responsibility rests with us human beings, who have used the wonderful powers put at our disposal by technology sometimes unwisely and even malignly, so that the question we have posed is one of human choice, and a satisfactory answer requires recognition of the need for some human changes of attitude. But the human responsibility is a collective one, in the sense that whole communities must recognise it. Individuals can do little to influence it, although they can urge their fellow citizens to join with them in accepting responsibility for the benign control of technology. As the old adage has it, a good workman does not blame his tools for the failure of his creations, and likewise any failure in the use of technology is the responsibility of those of us who use it. So the question can be rephrased as: can the world community devise a workable institutional framework which will prevent the malign use of technology and ensure that its creative value will be promoted?

The role of the state

The first response to such a requirement for the control of technology is the recognition that some form of viable world governance is essential for our survival. One of the functions of all state governments is that of exercising sovereignty – the defence of the community from external attack. Such government has historically taken many forms, but in recent centuries it is the form of the nation state that has been most widely adopted. Unhappily, this has created a situation in which rivalry between nation states has caused war, and with the growth in size of nation states and their increasingly powerful technologies of destruction, such a sequence of events cannot be allowed to happen again.

Since the end of the Second World War, and under the shadow of such horrors as atomic bombs and techniques of chemical and biological warfare, serious efforts have been made by the world community to reach beyond the inadequacies of the nation state by forming wider combinations such as the European Union and other regional associations, and also an all-embracing union in the shape of the United Nations Organisation. Despite much excellent co-operative work in trading relationships and in such fields as improving standards of world health and food supplies, these international endeavours have stopped short at handing over any national sovereign powers to the larger organisations. As expressions of groups with shared languages, cultures, and traditions, nations have enjoyed a long and rewarding history, but it is necessary, in the situation of endemic insecurity in which the world community now finds itself, to tease this idea of nationality away from its association with state sovereignty, and to vest such sovereignty in some form of world state. In the long run, only such an organisation can give any realistic hope of permanent security against the perils of conflict with the weapons of high technology, and the short run may prove to be very short indeed.

It is not necessary to be very prescriptive about the forms of this putative world state. A large measure of internal sovereignty can be left with the member nation states, as in the best-organised federal constitutions. They can retain traditional means of deciding between democratic, aristocratic, or monarchical forms of governance, even though modern experience has suggested that democracy is inherently the most stable form of society in the sense that it gives scope for every individual to express his or her talents creatively. It is thus one towards which member nations of a world community should be encouraged to develop. But the essential part of any arrangement of a satisfactory world order is that member nations must surrender to the international authority the role of defence, with a well-equipped militia, capable of intervening if necessary in any quarrel between members, replacing traditional national armed forces. Such an arrangement is not an idle pipe-dream, as many of the institutional prototypes required for its accomplishment are already in place in some of the regional organisations and in the UNO. What is needed is the vision and determination to make them function properly.

In addition to facilitating this fundamental function of the state to maintain peaceful conditions of life, technology has also become intimately involved in many aspects of the administration of nation states, such as legal arrangements for patent protection and the supervision of health and safety concerns in factories. Governments have a choice of options here, ranging from minimal intervention, in which everything except the maintenance of peace and law and order is left to the free decision of individuals concerned or to the operation of the market; and maximum intervention, in which all such arrangements are subject to close supervision by officers of the state. In practice, however, neither absolute choice is viable, and the skill of political life is devoted to finding the balance between them most appropriate for the state and for the time. Most modern states have found it necessary to make legal provision for the protection of innovations in patent law and, as far as possible, to arrange for interchanges with other national patent systems. They have also been obliged to provide inspection and supervision of potentially hazardous industrial processes and transport systems. The development of railways in the 1830s and 40s, for instance, required in Britain the creation of a small but effective inspectorate which was empowered to investigate all new lines before they were allowed to carry passengers, and these powers were subsequently extended to other aspects of railway operation. The alkali industry and gunpowder manufacture, amongst other processes, acquired similar government supervision, bringing a gradual increase of state control into new and old technologies and, in some cases such as the provision of pure water to households, to local or national ownership of public services.

In Britain such controls reached a high level after the Second World War, with gas, electricity, water supply, the railways, and leading industries such as iron and steel and coal mining, all brought into national ownership. Subsequent British governments have swung back to private ownership and market control for many of these industries, but a significant amount of state supervision of technological organisations remains in the control of local or national government, providing an important dimension to the regulation of increasingly complex technical systems. In addition, more socially directed intervention by the state such as factory legislation to control the work of children in factories and of women in coal mines, and in the development of a national system of education, has had significant technological implications for public policy. These, however, may be regarded more specifically as social history and treated as such.

Domestic technology

At the level of domestic life, technology has had a dramatic social effect in the period of High Technology, and particularly in the last hundred years. The basic feature in this transformation has been the provision of easily available power to the homes of the people in the shape of steam and internal combustion engines and, most recently, gas and electricity. The property of coal gas as an illuminant was discovered by William Murdoch, a colleague of Boulton and Watt, who illuminated their Birmingham factory in 1802 with coal gas which Murdoch derived from coal by heating it in a retort, driving off a gas that proved to be highly flammable. Within a few decades every British town of any size was equipped with a gas works supplying piped gas to factories and houses and for street lighting. The introduction of electricity to provide a similar service at the end of the nineteenth century stimulated the gas industry to improve its lighting qualities by burning it through gauze in 'gas mantles', which greatly improved the quality of illumination, but for sheer

Fig 10.2: New Lanark. Many successful entrepreneurs have shown care for the comfort of their workforce by providing special housing and public facilities, and one of the first to show such consideration was Robert Owen at New Lanark in Scotland in the early nineteenth century. (*Angus Buchanan*)

convenience it could not long compete with electricity in providing lighting. Gas, however, was adapted to changed circumstances by developing its uses in heating furnaces and ovens, and in this function it remains important today, even though the gas supplied is now normally 'natural' gas or gas produced from oil fuels.

Electricity has challenged gas in this field also, and seems likely to become the dominant form of energy in all forms of lighting and heating. Electric light was first produced as 'arc light' by passing a strong current between two adjacent rods, and was introduced in this form in lighthouses in the 1850s; but the discovery in the 1880s by Edison in America and Swan in Britain that a carbon filament placed in a vacuum in a glass bulb could be made to glow brightly by passing a current through it, made electric lighting available for domestic use, literally at the drop of a switch. The subsequent discovery that such light bulbs produced a weak 'free' current of electrons led to their widespread adoption for the 'valves' in wireless equipment, and soon after that to the invention of television, the computer, and much of the modern world's entertainment equipment. Electricity has thus had a tremendous transformative effect on modern life and culture. It has also provided power for transport systems and for many industrial applications, but it is in the home that it is most immediately of service to men and women, in lighting and heating, in powering electric ovens, washing machines, vacuum cleaners, and a bewildering range of household gadgets, in addition to the radios and television sets and electronic equipment which have come to figure so prominently in modern life and leisure.

Another domestic amenity of enormous importance, the provision of a reliable supply of pure water for drinking and cleaning, relies on much older techniques of water control by dams and reservoirs, aqueducts and filtration processes; but these required a measure of social organisation that was deficient in Britain until required to deal with the problems of supplying large industrial towns and cities. Only then did the resources become available to construct chains of dams impounding water on a large scale, such as those built in the mid-nineteenth century by the Manchester Corporation in the Peak District Longdendale valley. These were earth embankments with a core of puddled clay to make them watertight, and with an up-stream facing of masonry. Later dams tended to be high masonry or concrete walls like the Hoover Dam in Colorado, forming what was America's largest man-made reservoir when it was completed in 1935. Such dams are curved upstream to transfer the pressure of the water into the shoulders of the dam which are excavated into solid rock. The water retained in all such dams requires cleaning by passing it through filter beds before conveying it through to the citizens – sometimes many miles away.

At the other end of this process, that of disposing of water-borne waste or sewage, civil engineers have developed elaborate techniques of channelling the fluid under gravity through pipes and conduits, usually underground, to filter beds for purification before releasing it to natural water courses. Steam engines were regularly installed in the nineteenth century to pump the fluid to higher levels as required by the topography, but these have been almost entirely replaced by electric pumps. By these means rivers have been kept relatively clean, compared with the disgusting condition of urban rivers when used as the direct repository for waste of all sorts, such as was the case with the River Thames in London before the development of this particular technology.

Fig 10.3: Charterhouse-on-Mendip in, Somerset was worked over from at least Roman times to the early twentieth century, and the remains are slowly blending back into the landscape. (*Angus Buchanan*)

Chemistry has been another formative technology in the modern home, providing many novel materials such as 'plastics', mainly from a cellular base and taking many different forms, from celluloid for photographic film, to nylon – the wonderful new textile fabric of the Second World War – and bakelite and polythene, capable of being moulded in many hard forms, which have revolutionised the ceramics and furnishing industries and, through the production of celluloid, made possible the popular adoption of camera photography. The camera itself – a box allowing light to be focussed through a pin-hole or glass lens – had been adapted to catch sunlit images on metal, glass, or paper, which had been coated with solutions of light-sensitive silver oxides from which they could be 'developed' by chemical fixatives in a 'dark room'. But it was the use of a celluloid film as his base by the American George Eastman in his 'Kodak' camera of 1888 which made photography available to anybody with modest financial resources. Then in 1894 another American,

Fig 10.4: Crystal Palace. The superb glass and iron structure built in Hyde Park to house the Great Exhibition of 1851 contributed significantly to the success of this event. The Palace was subsequently dismantled and re-erected at Croydon, where it was destroyed by fire in 1936. (*Artist Unknown*)

Thomas Edison, introduced narrow strips of celluloid as the basis for his 'movie' film with perforations which passed through toothed wheels to give a sequence of images reproducing the appearance of movement. Edison immediately appreciated the popular application of this invention by establishing the first 'cinema' in which the public could pay to attend demonstrations of short films, thus giving birth to one of the greatest modern entertainment industries. In all these ways, technology has had and continues to have a profound influence on the way we live in modern society – and beyond, in the wider environment.

The environment

Once the bane of conflict between nation states is removed, with the important technological aids of inter-communication and consultation, the new organisation of world governance can turn with hope of real success to international problems of the environment. These include the response to climate change, the depletion of essential resources such as drinking water and rare metals, the provision of good quality food for the world population, the safeguarding of whales and the fish stocks of the world's oceans, the maintenance of ecological homes for the world wildlife, and stabilising the human population of the world at a level that can be accommodated comfortably. In all these fields technology has an important continuing role to play, made all the more powerful by being released from the imperatives of arms production, which will be achieved by elimination of the fear of war.

Anxiety about environmental problems has been prevalent for a long time, but they were brought into sharp focus after the end of the Second World War by the publication in 1962 of *Silent Spring* by the American scientist Rachel Carson. By drawing attention to the loss of song birds in American suburban gardens, she exposed the destruction of wildlife through the widespread and indiscriminate use of pesticides, and the dangers to human life implicit in the deposition of such poisonous substances in the food chain. The chemical and pharmaceutical industries responsible for perpetrating these hazardous conditions dismissed Rachel Carson as a scaremonger, but within a few years the thrust of her argument was confirmed by chilling events such as the discovery of the disastrous consequences of using thalidomide as a medical drug, and her views were rapidly assimilated. Some of the most sinister chemicals were banned altogether, while others were put under much closer control.

Perhaps of even greater significance than the immediate effects of Rachel Carson's message were its long-term consequences, because she stimulated a great increase of interest in ecology – the science of understanding the balance and interconnection between the many species sharing a home on Planet Earth, with each species requiring an 'ecological niche' within which it could live and procreate. This ecological consciousness spawned a large literature as people with many different concerns and points of view contributed their research, which reached a culmination with the development of the *Gaia* concept by James Lovelock in 1972. Taking its name from the Greek Goddess of the Earth, Lovelock sees Gaia as the biosphere of Planet Earth – the thin covering of earth, sea and atmosphere surrounding the planet, the parts and inhabitants of which interact with each other to sustain a stable, self-regulating, environment. But, he argues, 'by changing the environment we have unknowingly declared war on Gaia' [*Revenge*, p.10]. In particular, by burning fossil fuels and pouring carbon dioxide into the atmosphere, while destroying rain forest and other effective modes of absorbing the excess CO_2, we have initiated a rapid process of global warming which is causing climate change and rising ocean levels with consequences which we are only just beginning to recognise and take measures to correct. Lovelock ends up in a rather melancholic and doom-laden position of believing that we have left corrective measures almost too late to save the human species, but it would not be human nature to accept such a conclusion, and provided that we act purposefully and promptly there are still grounds for hope that we can take steps towards restoring the naturally self-controlling planetary system of Gaia.

Such steps include making changes in our cultural attitudes, so that we recognise the crucial role of science and technology in determining the structure and future of our society. This is primarily a matter of modifying our attitude towards education, not just as a scholarly or academic concern but regarding the continuing need for individuals to increase their knowledge and understanding of the world around us. Such a profound reorientation of human attitudes would bring with it a willingness to adopt new goals, such as those already mentioned – increasing the world's food supply and conserving its precious water, improving conditions of health and hygiene amongst the population of the world, cleaning up the environment and husbanding the non-renewable resources of the planet, and securing sympathetic policies of population control. All this would encourage an aspiration to promote the systematic exploration of the universe, made possible by being released from the imperatives of arms production and the elimination of the fear of war.

The objectives of a technological society

If the world community can begin to achieve these goals – and it must remain an 'if' because we have already left the solution to some of the problems such as dealing with climate change and atmospheric pollution dangerously late – it will be able to undertake a programme of cultural enrichment. This should include improving standards of education throughout the world, so that the power of technology can be more generally understood, together with the continuing need to control it and to use it creatively. It should also contain a vision of future technological possibilities, including that of expanding the prospect of space exploration which developed so excitingly in the second half of the twentieth century but has been forced onto the metaphorical 'back-burner' by the preoccupations of the world community

Fig 10.5: Earth-rise over the Moon. This iconic photograph was one of a series taken from the Apollo space ships as they emerged from orbit round the Moon to see the Earth rising before them. (*NASA*)

with more immediately pressing problems of world organisation. This possibility has hitherto been subverted by military and other demands, so that it is taking longer to fulfil the initiatives already established in Lunar and Martian exploration than seemed possible half a generation ago. But the unmanned missions to the outer solar system and to deep space beyond continue to bring discoveries and surprises, as science and technology steadily advance our power to probe the nature of the universe.

Even after such a general review of the relationships between technology and society as can be attempted here, it should be pretty clear that modern society has been profoundly influenced by technology and that this influence continues with no sign of abatement. From empowering national governments to control and maintain law and order through instant communication and instruments of internal discipline, to wielding power to affect the environment through climate change by reversing deforestation and desertification, with all the intermediate processes of feeding the population and keeping it well heated and lighted and entertained by ubiquitous electronic equipment, technology has permeated every level of social organisation and become indispensable to its survival. It also dominates our future prospects in population control and cosmic expectations, and in one respect – that of warfare – it continues to pose a threat to the survival of civilised society.

Further reading

Buchanan, R A: *The Power of the Machine,* (Viking Penguin, London, 1992).

Carson, Rachel: *Silent Sprin,* (1962, Penguin Modern Classics, 2000).

Clarke, Arthur C: *The Exploration of Space,* (Penguin/Pelican, London, 1958).

Clarke, Arthur C: *Profiles of the Future,* (Pan Books, London, 1958).

Lovelock, James: *Homage to Gaia: The Life of an Independent Scientist,* (Oxford UP, 2000).

Lovelock, James: *The Revenge of Gaia: Why the Earth is Fighting Back – and How We Can Still Save Humanity,* (Allen Lane/Penguin, 2006).

CHAPTER 10.1

Postscript: Warfare and Society

Brenda Buchanan

Warfare has long been a subject of study by historians, some interested chiefly in weapons and tactics and others concerned mainly with the relationship of war to the fortunes of states and society. In most cases the provision of firepower has been seen as a 'given factor', something that was essential but which could be taken largely for granted, barely worth a footnote until studies focussing on the manufacture of gunpowder were published from the later decades of the twentieth century. These suggest that in the western world the gradual adoption of gunpowder weapons from the fourteenth century was based upon mechanical mixing procedures that appear simple but were only gradually understood by contemporary society, in contrast to the science-based firepower adopted from the mid-nineteenth century that was developed through chemical explosives laboratories and factory floor industries, revealing a society able to more rapidly develop and adopt new means of warfare.

In the international context the scholarly research of Joseph Needham, published in the 1980s, had already shown the ninth century primacy of gunpowder knowledge in China, and to this may be added the suggested twelfth–thirteenth century use there of this energetic power for warfare and pleasure, explored in the historical research of Kenneth Chase and Peter Lorge, published in the first decade of the twenty-first century. The significance of this widening of focus is such that western historians must now take into account not only the professional disputes within their own geographical area of research, but also the possibility that in chronological terms the gunpowder armies of the west did not precede those of the east. And perhaps in any case the long debate on the idea of a Military Revolution in western civilisation has now run its course, having engaged much attention in the second half of the twentieth century after Michael Roberts' revision to our understanding of warfare and national administration between 1560 and 1660, first made in 1956. Others have since accepted the concept but questioned the chronology, preferring the period after 1660 as being of greater importance for military innovations. Such revisions indicate the need for a more encompassing context and chronology within which to view the changes in warfare implicit in the term 'military revolution', incorporating in these considerations the less contentious but no less dramatic and far-reaching developments of the second half of the nineteenth century already noted.

In light of these considerations it may be suggested that the development of military skills in the west occurred gradually over many centuries, but accelerated with dramatic effect with two technological innovations during which new sources of destructive power became available to armed forces. The first, in the fourteenth–fifteenth century, was the introduction of gunpowder weaponry and the second, in the mid-nineteenth century, came with the application of high explosives and mechanisation to warfare. Both of these developments were associated with the growth of the nation state and the size of armies, with social consequences in terms of funding, recruitment, and logistics.

A knowledge of gunpowder and the fundamental and powerful changes in warfare to which this was gradually to lead came, as noted above, not from within the medieval society of western civilisation, but from mandarin China where experimental research was encouraged as part of the quest for the ultimately unattainable elixir of life. As the Taoist alchemists collected, refined, and experimented with various ingredients in simple laboratories, it was only a matter of time before in the ninth century the three main constituents of gunpowder – saltpetre, sulphur and charcoal – were mixed together and inflamed. The results were startling, with the fire, smoke and noise that were initially used to entertain the Court of the Emperor becoming a significant aspect of warfare, with the firing and explosions of bombs, guns and rockets. This knowledge was carried to the west as part of the diffusion of Chinese learning in the twelfth and thirteenth centuries, perhaps along the routes later to become known as the Silk Road. It had reached Paris by at least the mid-thirteenth century, when Roger Bacon of Oxford became the first western scholar to write about the flashes of light and roars like thunder of exploding firecrackers, expressing concern about the 'greater horrors' that might lie ahead. Bacon's writings are approached with caution by Bert S Hall in his 1999 introduction to J R Partington's study of 1960, but despite concerns about the deciphering of codes, gunpowder was becoming known.

The adoption of gunpowder in the west was to have over time a profound effect upon both warfare and civil society. The ingredients of gunpowder remained simple, but skilful experimenters sought the preparations and combinations that could produce the 'best' mixture for the use intended. The skills of the metallurgist were employed in devising efficient handguns and artillery. The skills of the engineer were valued as the designs of castles and other fortifications underwent fundamental changes in order to provide better protection against the new gunpowder weapons, especially heavy cannons. These developments included the construction of projecting towers and angled bastions, devised to provide cross-fire in a system adopted in the late fifteenth and early sixteenth centuries by Italian city states and known for that reason as *trace-italienne*. The impact of gunpowder on civil society was also profound although its adoption in quarrying, engineering, mining and trade came later, and cannot be explored in this short study. It was used from the mid-fifteenth century by military engineers to undermine fortifications, level ground and deepen rivers, but it was not until the early seventeenth century that powder was used successfully for blasting in silver, lead and copper mines in central Europe, and not until a century later was it used in coal mines in England for example, for sinking shafts. A failed attempt at Schio in northern Italy in the 1570s, on a site belonging to the state of Venice, might have led to an earlier introduction of black/gunpowder, since contemporary State Papers refer to its use to break up rock to determine its silver-bearing qualities. Other investigations followed, perhaps on documentary evidence most successfully at copper mines in the Le Thillot district of the Vosges, worked for two centuries from as early as 1617, but the successful introduction of shotfiring at Schemenitz (then in Hungary) was the first to name the successful mining engineer concerned, Kaspar Weindl. It was perhaps the skills of such experts in moderating the practices of military engineers in order to achieve a more controlled use of powder that facilitated its civil in addition to its military use.

In terms of its military use, the impact of gunpowder weapons on the order of society was very great, and there was at first some social confusion as to who was to have the responsibility for exercising this new physical power. Illustrations in a manuscript compiled by Walter de Milemete 1326/27, *Concerning the Majesty, Wisdom and Prudence of Kings,* provide us with the first picture of a cannon in Europe, one that is intriguingly similar in its pear or vase shape to a Chinese illustration of c.1300.

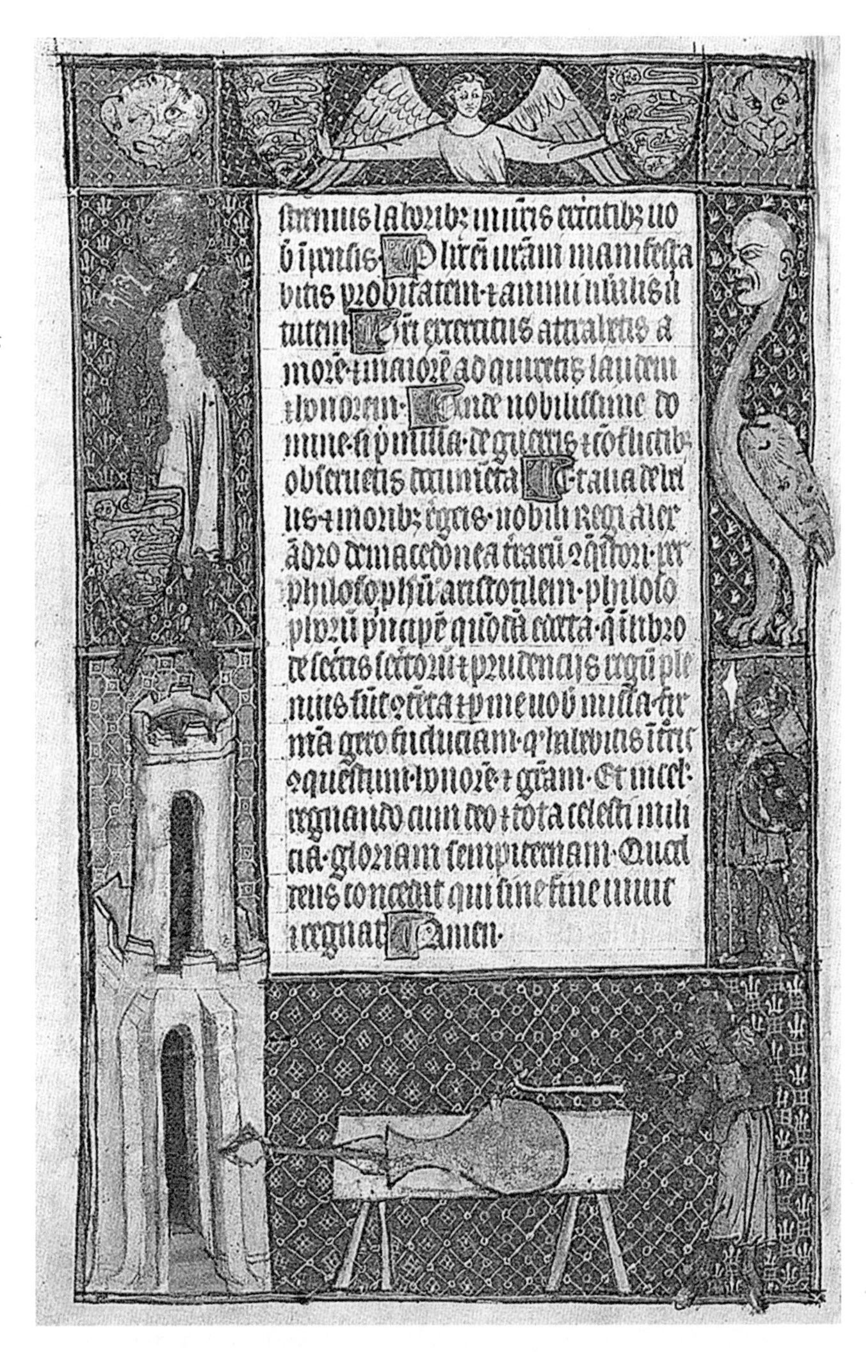

Fig 10.1.1: Walter de Milemete's fourteenth century manuscript provides our first illustration of a gunpowder weapon (see text). (*From Christ Church Library, Oxford, Western Manuscripts Collection, MS 92, folio 70v. This image from De Nobilitatibus, Sapientiis, et Prudentiis Regum of 1326/7 is reproduced by permission of the Governing Body of Christ Church, Oxford*)

The Milemete illustration shown in Figure 10.1.1 also suggests incidentally the difficulty presented to an established social order by such a major innovation in the weapons of war. Knightly figures in armour covered by loose surcoats and not lower-ranking soldiers are shown, one firing a vase-shaped cannon resting on its side on a simple bench with an arrow targeted at a stone castle about to fly from its muzzle. The death of James II of Scotland in 1460, blown up whilst standing too close to a cannon that shattered on being fired, shows that some were tempted to be neither wise nor prudent in these matters. But with the development of the professional skills of the artillerymen, and the equipping of foot-soldiers with portable handguns, distinct and separate skills were promoted within armies that allowed officers to retain their traditional position, wielding swords from their superior height on horseback.

Securing good supplies of the three essential ingredients of gunpowder became crucial to the conduct of this new form of warfare in the west. All were first produced from available domestic sources, with the acquisition of saltpetre (potassium nitrate) especially arousing strong resentment because it was derived from 'black earth', nitrogenous waste collected from animal and human middens by scavenging saltpetremen. Sulphur could be produced by chemical processes, especially the roasting of pyrites, but with the development of Mediterranean trade it became more accessible from volcanic sources in Italy and Sicily. Charcoal was fairly easily available from woodland sites, but care had to be taken in choosing the best wood for the purpose and in charring it correctly, excluding air as much as possible. However, with the growing demand for gunpowder in civil engineering, mining and trade (for example for the 'purchase' of slaves by barter), as well as for military purposes, it was important that new sources of the major ingredient, saltpetre, should be found.

The search for saltpetre had significant consequences politically as well as militarily. The unsavoury activities of the saltpetremen caused such resentment within society that the possibility of trade with India in this commodity was attractive. By the mid-seventeenth century the initial stage of trial and error by the English East India Company was coming to an end and from the early eighteenth century growing cargoes of partially processed saltpetre were being shipped to Britain, where it was further refined into a high grade commodity. Although several other western European countries set up their own general trading companies, from the mid-eighteenth century this trade was largely in the hands of the English East India Company, giving a vital political as well as military advantage to Britain. As with India, access to supplies of sulphur from Sicily in particular provided a reason for the British strategic interest in this region. Charcoal continued to be available from domestic sources in Britain but to maintain supplies, especially of the well-regarded alder and alder buckthorn, plantations were sometimes set up at powder mills where trees were pollarded to maintain supplies.

When brought together in their correct proportions, which varied according to time and place but came generally to consist for military use of 75 per cent saltpetre, 10 per cent sulphur, and 15 per cent charcoal, the ingredients were mixed under pressure. They did not undergo a chemical combination but were instead subjected to a mechanical mixing, using powerful descending stamps or heavy edge-running millstones. This mechanical incorporation had to be so thorough that when the dampened mix was sieved to produce pellets, each should contain the ingredients in their correct proportions. With some refinements this relatively simple process produced adequate fire-power for British military and naval purposes until developments from the middle of the nineteenth century began to make available chemically-based high and low explosives, propellants and detonators, for use in rifled muskets and breech-loading heavy ordnance.

For reasons of security, the governments of European countries made great efforts to control the production of gunpowder by private individuals and partnerships, whilst valuing their significant role in supplementing military supplies when required, and in developing profitable trades. But whereas some, such as the Portuguese state authorities, had established early monopolies of production that were to lead to the setting up in 1725 of the Barcarena Royal Gunpowder Factory outside Lisbon, the British government continued to rely on private producers until in 1787 the Waltham Abbey Gunpowder Works were purchased. These were located conveniently for London on a tributary of the Thames, the River Lea, which provided both transport and water-power, with steam power being introduced at the now 'Royal' factory in the mid-nineteenth century. Gunpowder production continued but the manufacture of cellulose-based chemical propellants, guncotton and cordite, was developed here from the 1860s, and some of the gunpowder produced was for specialised use as an igniter or fuse for the smokeless explosive cordite that was to become the principal British service propellant. Other specialised products were the pellet, pebble and prismatic cannon powders for heavy guns. With a developing emphasis on research, the Waltham Abbey RGF was to continue after the Second World War as a major centre for work on non-nuclear explosives and propellants, though this had ceased when the site was eventually closed by the Ministry of Defence in 1991.

In close connection with the rise of nation states capable of raising and sustaining large armies, the versatility of gunpowder in its application to new forms of weaponry had brought changes in warfare between the fifteenth and nineteenth centuries. But this was a comparatively 'slow burn' revolution compared to that associated with the exploitation of the new high explosives and mechanical resources of highly industrialised societies, which produced a second 'step-change' in the 'military revolution' from the middle of the nineteenth century. The discovery of the explosive qualities of nitrocellulose and nitroglycerin by chemical scientists brought into existence a succession of explosives of much greater power than gunpowder, and these were rapidly adopted for use in small arms, for which they had the advantage of causing less smoke than gunpowder and enabling greater range. Larger and stronger guns were made to propel shells with great powers of penetration and devastation. They were also readily adapted for use in grenades, bombs and rockets. Taken together, they transformed military and

Fig 10.1.2: A 'Gunpowder Salute' by the author, firing a replica of a Falconet of c.1650 on a field carriage, the property of the Old Wardour Castle Garrison. The effective range of c.500 yards was increased to c.1800 yards if shot at the 'utmost random'. I am indebted to Dr Karel Zeithammer of Prague for this photograph, and to Mr W Curtis for information on the Falconet.

naval tactics and the scale of devastation caused by armed conflict in the decades culminating in the First World War.

As the introduction of high explosives coincided with the increased mechanisation of industrial society, warfare itself assumed a highly mechanised character, with the machine gun, the tank, motorised infantry, the aeroplane, and radio and telecommunications, all being introduced into the arsenals available to modern military combatants. The transformation of the sailing ship into the steam-propelled ironclad vessel provided another platform for the new explosives, with high powered guns firing shells capable of greater penetration than traditional cannon. Poison gas and bio-chemical weapons have also been produced by modern science – although these have been banned by international authorities. Such bans however are never completely reliable and the threat these weapons pose, and the still greater horrors of a conflict with atomic weapons, present the world community with its greatest challenge – that of instant and total destruction at the press of a few buttons. The threat of warfare on this scale represents the ultimate stage in the long process of military revolution and has become the greatest challenge facing the survival of society, although the task of containing and preventing acts of individual terror is a further continuing requirement for all governments.

Further reading

There are many texts on this subject of which the following are a sample:

Black, Jeremy: *European Warfare 1660–1815,* (Yale University Press, 1994).

Buchanan, Brenda J, ed.: *Gunpowder: The History of an International Technology,* (Bath University Press, 1996, reprinted 2006).

Buchanan, Brenda J: 'The Art and Mystery of Making Gunpowder: The English Experience in the Seventeenth and Eighteenth Centuries' in Brett D Steele and Tamera Dorland, eds., *The Heirs of Archimedes. Science and the Art of War through the Age of Enlightenment,* (The MIT Press, Cambridge Mass., 2005), pp. 233–274.

Buchanan, Brenda J: 'Saltpetre: A Commodity of Empire' in Brenda J Buchanan ed., *Gunpowder, Explosives and the State: A Technological History,* (Ashgate, Aldershot, 2006), pp.67–90.

Chase, Kenneth: *Firearms: A Global History to 1700,* (Cambridge University Press, 2003).

Cocroft, Wayne D: *Dangerous Energy. The Archaeology of Gunpowder and Military Explosives Manufacture,* (English Heritage at the National Monuments Record Centre, 2000).

Hall, Bert S: *A New Introduction to J R Partington's History of Greek Fire and Gunpowder* 1960, (The John Hopkins University Press, 1999).

Lorge, Peter A: *The Asian Military Revolution. From Gunpowder to the Bomb,* (Cambridge University Press, 2008).

Needham, Joseph et al: Science and Civilisation in China. Vol. 5, Part 7, *Chemistry and Chemical Technology: Military Technology, The Gunpowder Epic,* (Cambridge University Press, 1986).

Parker, Geoffrey: *The Military Revolution: Military Innovation and the Rise of the West, 1500–1800,* (Cambridge University Press, 1988).

Parker, Geoffrey: *The Cambridge History of Warfare,* (Cambridge University Press, rev.ed. 2008).

Roberts, Michael: *The Military Revolution 1560–1660,* (University of Belfast, 1956).

Chapter 11

Technological Prospects

David Ashford

Predicting developments in technology is an uncertain business. History shows many examples of inventions that were thought to be imminent but which have never happened. Conversely, several developments have taken most of us by surprise. For example, did the inventors of the computer, or science fiction writers, or anyone else for that matter, predict the Internet, spam, hacking, viruses, or cyber warfare?

In spite of this difficulty, two future developments are considered in this chapter as being worth taken seriously. The first concerns spaceflight, for which an imminent revolution can be predicted with some certainty. The second, longer term and more speculative, is beginning to be taken seriously and is potentially so important that it is worth considering some of the possibilities. This concerns the prospect of developments in medicine and computing combining to affect our very nature.

Starting with spaceflight, an imminent revolution can be predicted reliably because the obstacle to its happening is not the technology, the market, the economics, or the politics – it is no more than, and no less than, mindset; and enough work is now in progress, led by the private sector, to make the overturn of this mindset all but inevitable. At present, the major players, who are mainly large government space agencies and their contractors, just do not want to know. It is an incredible situation in which an obvious and highly desirable development has been suppressed not by conspiracy (overt or covert) but by corporate groupthink. What is now the most competitive way ahead was widely considered feasible in the 1960s.

The revolution involves replacing today's throwaway launchers with those like aeroplanes (spaceplanes) based on designs widely studied in the 1960s but using up-to-date (but not advanced) technology. These offer greatly reduced cost and improved safety and can provide an airline service to orbit.

The primary cause of the high cost and risk of spaceflight is that launch vehicles to date have used components based on ballistic missile technology that can fly only once. Imagine how much motoring would cost if cars were scrapped after each journey! The revolution involves replacing these missile-like launchers with ones like aeroplanes that can fly to space many times (spaceplanes).

Figure 11.1 shows a typical expendable launcher, Ariane 5, and the Space Shuttle. Ariane 5 is entirely expendable. After each launch, the various stages either crash into the sea or burn up on re-entering the atmosphere. The Orbiter stage of the Space Shuttle

Fig 11.1a and 11.1b: Ariane 5 and the Space Shuttle. (*ESA and NASA*)

Fig 11.1b

was like an aeroplane and was reusable. The largest component was the External Tank, and this burnt up on re-entry. The two Solid Rocket Boosters were recovered at sea by parachute. They were then returned to the factory where some of the components were refurbished and used again. From the point of view of cost and safety, they were little better than if they had been expendable.

Ariane 5 and the Space Shuttle launch(ed) spacecraft into orbit. This requires climbing clear of the effective atmosphere to space height and then accelerating horizontally to satellite speed, which is about 17,500mph (7.8km/sec) or round the world in ninety minutes. It is the acceleration to satellite speed that requires most of the energy – just climbing to space height requires far less. A so-called suborbital flight – up and down to space height with just a few minutes in space – requires a maximum speed of about 2,200mph (1km/sec), as shown in Figure 11.2.

There is no clear boundary between the atmosphere and space, but space height is usually defined as 62 miles (100km). Expendable suborbital rockets, called sounding rockets, have been used for many years for various kinds of scientific research. They are far smaller and less expensive than orbital launchers, but only provide a few minutes in space. As will be discussed later, two

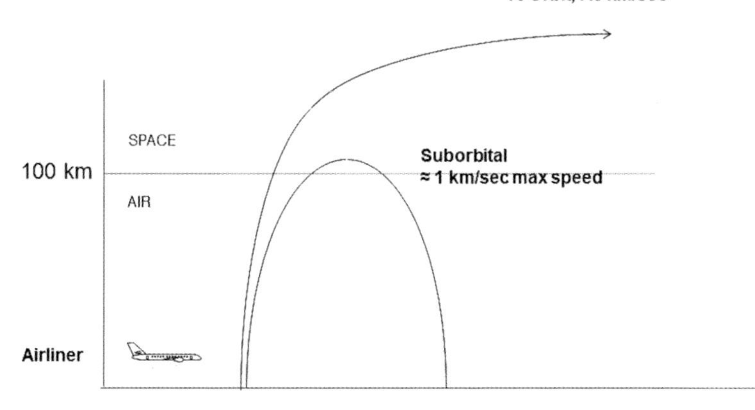

Fig 11.2: Orbital and suborbital launch trajectories. (*Bristol Spaceplanes*)

fully reusable suborbital spaceplanes have already flown, and new ones are being developed to provide passengers with suborbital space experience flights. These will be useful stepping-stones to orbital spaceplanes.

How has this persistent use of throwaway launchers come about? The answer lies in the history of spaceflight. The first spacefaring vehicle was the German V-2 ballistic missile of the Second World War, which first reached space height in 1942 (see Figure 11.3). The V-2 was a suborbital vehicle, spending little time in space before re-entering the atmosphere and delivering its warhead. It had the now-classic configuration, with the payload, in this case a high-explosive warhead, at the front. Most of the body was filled with rocket propellant, in this instance liquid oxygen as the oxidizer and alcohol diluted with water as the fuel. The rocket engine was at the base. This was where the oxidizer and fuel reacted to produce a hot gas at high pressure, which then emerged at high speed through the nozzle to produce the thrust.

Fig 11.3: The German V-2 ballistic missile of the Second World War. (*Deutsches Museum*)

More than 3,000 V-2s were launched, mostly against Antwerp and London, in the last year of the Second World War. It was a brilliant weapon technically but it was not very effective because of poor reliability and accuracy. The only target that it stood much chance of hitting was a large city and even there it was unlikely to damage more than a few buildings.

The nature of its task meant that the V-2 was expendable, although a piloted reusable bomber version, with wings for landing, was designed. However, the war ended before this version could be built.

The V-2 was years ahead of any competition, and formed the basis for post-war US and Soviet ballistic missile development. Their aim was to overcome the limitations of the V-2 by combining more accurate guidance systems with nuclear warheads. The first indigenous US ballistic missile, the Redstone, was in effect a re-engineered and enlarged V-2. Its design team was led by Germans recruited to work in the US, the best known being Wernher von Braun, who had been the prime mover behind the V-2.

Because ballistic missiles can fly to space, it was natural to use them to launch the first satellites. Redstone formed the lower stage of the four-stage *Juno 1*, which in 1958 became the first US vehicle to launch a satellite into orbit. This was in rapid response to the first ever satellite – the Soviet *Sputnik* launched in 1957. In 1961, Redstone also launched the first American into space, Alan Shepard, again in response to a Soviet first when Yuri Gagarin became the first man in space earlier that year. There was actually a big difference between these two achievements – Shepard's flight was suborbital whereas Gagarin's was fully orbital. Achievement in space became a key part in the propaganda battle between the Soviet Union and the US during the Cold War.

The next major step after sending men to orbit was the race to the Moon. The US effort was kicked off by President Kennedy's famous speech In May 1961, just six weeks after Gagarin's flight to space, when he galvanized the nation by saying:

> *'If we are to win the battle that is going on around the world between freedom and tyranny, if we are to win the battle for men's minds … I believe that this nation should commit itself to achieving the goal, before this decade is out, of landing a man on the Moon and returning him safely to Earth.'*

The first sentence in the quotation shows how political the objectives were. Kennedy was far more interested in beating the Soviets than in exploring space for its own sake. Project Apollo was a brilliant success: twelve men stood on the Moon between 1969 and 1972, and all returned safely. The Soviet effort was an expensive failure. The West went on to win the Cold War, Apollo having played its part.

To save time in the race to the Moon, the mighty Saturn series of large new launchers was built as expendable. Von Braun had earlier proposed a large reusable vehicle, using the experience of two experimental V-2s that had flown with wings in 1945, but there was not enough time to develop these and to be reasonably sure of reaching the Moon before the Soviets.

In parallel with the development of larger and more capable satellite launchers, the US led the way in developing a series of rocket-powered high-speed research aeroplanes. The Bell X-1 became the first piloted aeroplane to exceed the speed of sound (Mach 1) in 1947; the Douglas Skyrocket reached Mach 2 in 1953; and the Bell X-3 reached Mach 3 in 1956. The last in this series was the North American Aviation Inc. X-15 (see Figure 11.4), which made its first flight in 1959 and its last in 1968. The X-15 was designed to explore hypersonic speeds and it eventually reached Mach 6.7. Perhaps more importantly, it was also designed to bring an aviation approach to spaceflight by using rocket-powered aeroplanes instead of expendable converted ballistic missiles. Many of the X-15 team were driven by a vision of everyday economical transport to space. The X-15 could reach space height and was the first true suborbital spaceplane. It could have been used as the reusable lower stage of a satellite launcher, but a feasible proposal to do this was rejected due to concerns over safety and economics. As well as carrying out invaluable aeronautical research, the X-15 carried several space science experiments, serving as a reusable sounding rocket.

Jumping ahead in our story, the Scaled Composites *SpaceShipOne* (SS1), also shown in Figure 11.4, was the next aeroplane to reach space height. It achieved this feat in 2004, thirty-six years after the X-15 last flew to space. This hiatus is an indication of the lack of priority given to reducing the cost of access to space.

In parallel with Apollo development, and partly inspired by the X-15, most large aircraft companies in Europe and the USA worked

Fig 11.4a and 114b: The X-15 and SpaceShipOne (SS1 – the only suborbital spaceplanes to have flown. (*NASA and Scaled Composites*)

up proposals for fully reusable launchers that could fly all the way to orbit, which would now be called orbital spaceplanes. At a Society of Automotive Engineers (SAE) space technology conference in Palo Alto, California, in 1967, for example, which I attended, no fewer than fifteen proposals for reusable launch vehicles were presented, and there were several others not represented there.

Among these were eight designs prepared by European companies as part of the so-called Aerospace Transporter project, some of which are shown in Figure 11.5. This programme was coordinated by Eurospace under the guiding mind of Professor Eugen Sänger, who had designed a rocket-powered suborbital bomber in the Second World War and who can well be described as the father of the spaceplane. These spaceplane designs were all relatively small, with a payload of one or two tonnes to orbit. Most of the US designs were considerably larger. The author's first job was working on the Hawker Siddeley Aviation (HSA) design.

There was a consensus that reusable launchers were the obvious next step in space transportation. Consuming a complete vehicle for each launch could never be economical. There was also a consensus that such vehicles were feasible with the technology of the day. The suborbital X-15 was demonstrating that the technical problems of orbital spaceplanes were within sight of being solved.

With just a few dissenters, there was a further consensus on the essential design features, and most of these 1960s projects (and all of the European ones) had the following in common:

- Two stages, a booster and an orbiter, so that the technology of the time could be used.
- The use of recently developed hydrogen-fuelled rockets, which had significantly improved performance.
- Wings to provide lift for landing, as being safer and more practicable than rotors, vertical jets, vertical rockets, or parachutes.
- Pilots, as being safer than autopilots or remote control.

Incredibly, any one of these projects would have been far better, in terms of reducing launch cost, than anything built since then, or even proposed by a major player. As will be discussed later, this set of key design features is still by far the most competitive.

The biggest unknowns at the time concerning spaceplane design were the effects of wings on re-entry stability and heating, and how

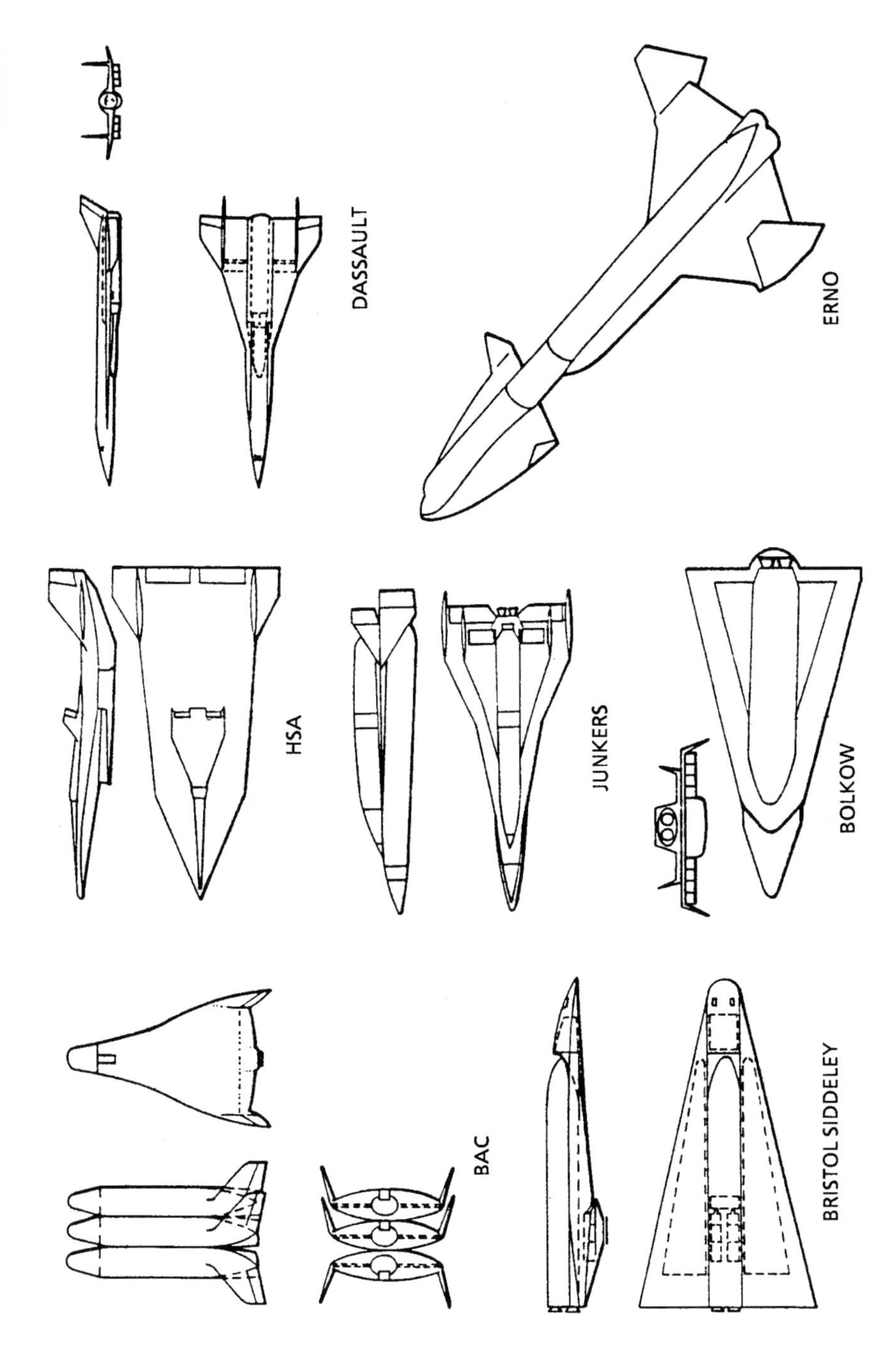

Fig 11.5: European Aerospace Transporter (spaceplane) projects from the mid-1960s. (*Bristol Spaceplanes*)

to protect the structure from the heat of re-entry. The Space Shuttle has since provided a full-scale demonstration of solutions to these problems, and there have been major advances in these and other technologies since the Shuttle was designed. All the technologies for an orbital spaceplane have now been demonstrated in flight.

Most of these 1960s spaceplane projects were designed mainly for launching satellites and sending people and supplies to space stations. However, they could have been adapted for carrying passengers on long-distance flights or for space tourism. If a spaceplane accelerates to just short of satellite speed, it can then glide halfway round the world. The flying time from Europe to Australia, for example, would be about seventy-five minutes. At the time, space tourism had hardly been considered seriously, and fast long-range transport received more attention. A few designers thought that this would be by far the largest market for spaceplanes and that the resulting economies of scale would reduce the cost per seat to just a few times more than the fare in a conventional airliner. This raised the prospect of a fleet of a few dozen spaceplanes each making several flights per day, leading in turn to airline-like operations. The result would be an aviation approach to space transportation to replace the missile one. These studies were the first to indicate the low-cost potential of spaceplanes when used in significant numbers. One of the first research reports on these lines was *The Rocket-propelled Commercial Airliner* by Walter Dornberger, published in 1956. This was the year before the first satellite launch. Earlier, as a Major-General in the German Army, Dornberger had led the V-2 programme.

Contemporary designers generally accepted this view as an interesting long-term prospect rather than as an immediate priority. However, if a spaceplane had been built in the 1960s for launching satellites, it would no doubt have been adapted for experimental long-range transport and for pioneering space tourism – carrying passengers to and from space hotels equipped to show the superb views of Earth and outer space, and fitted with a large zero-g gym. If there had turned out to be a large market for either of these uses, it is likely that regular commercial operations would have started by the late 1970s. Now, most consider that the first large market for spaceplanes is more likely to be space tourism. People will pay more for a once-in-a-lifetime visit to space than for a fast flight from Europe to Australia, and so a less economical vehicle can be used. Space tourism can therefore use a less advanced vehicle than is needed for fast long-range transport.

These three strands of development – improved expendable launchers, higher performing rocket-powered research aeroplanes, and spaceplane studies – came together for the next major project after Apollo, which was the Space Shuttle. When Shuttle design was started in the 1970s, the early proposals were fully reusable, taking advantage of X-15 experience and the numerous 1960s spaceplane studies. The project was so promising that there was great optimism about the future of space. Gerard O'Neill even proposed building space colonies in orbit, using materials from the Moon and asteroids, and these ideas gained much publicity. This was shortly after the film *2001: A Space Odyssey* (1968), which added to the general optimism.

The early designs for the Shuttle were very large, having a 30-tonne payload to meet the requirements of the US Department of Defense. Then President Richard Nixon cut NASA's budget, and the large fully reusable design could no longer be afforded. NASA then faced a choice. They could either make the vehicle far smaller but still fully reusable, along the lines of earlier projects (especially the European Aerospace Transporters), or they could give up on complete reusability. The habit of expendability was by then strong enough for NASA to make the great mistake of choosing the latter course.

As a result, the Space Shuttle was just as expensive to fly in terms of cost per tonne of payload as the Saturn that preceded it. This put paid to several promising schemes for exploiting low-cost space transportation, although it took several years for the limitations of the Shuttle to become widely appreciated. Pilot schemes for O'Neill's space colonies, for example, might just have been affordable with the reusable Shuttle originally planned but did not stand a chance with the largely expendable one. At the time of the Shuttle decision, NASA did not seem to appreciate the profound difference between an expendable and a reusable design. The former is fundamentally unsuitable for large routine markets; the latter can be developed to service them, like aeroplanes service the air transport market.

In spite of this, the Shuttle was the mainstay of the US human spaceflight programme for thirty years, from 1981 to 2011, and did much useful work. However, the decision to make it not fully reusable has delayed low-cost access to space by thirty years and counting. This history has created institutions and habits of thought that repeatedly reinforce the expendable habit. A great opportunity was missed to bring in an aviation culture to replace the missile one. Even today, most space agencies are promoting new expendable launchers.

How well do present space plans reflect the promise of these 1960s spaceplane studies? None of the space agencies is showing much interest in spaceplanes although NASA and the UK Space Agency are helping to fund some private sector initiatives. NASA's main preoccupation regarding space transportation is finding the budgets needed to build a very large new expendable launcher, SLS (Space Launch System), which, when fully developed, would be larger than the mighty Saturn of the 1960s.

There seems to be no rational reason for NASA and other space agencies to be so reluctant to consider seriously the prospects for spaceplanes. The entrenched habits of thought mentioned above are one explanation. A second explanation is that a cultural gap has grown up between the space and aviation industries. All of the European and many of the US spaceplane projects of the 1960s were carried out by *aeroplane* design teams, with perhaps some input from engineers with experience of launchers or re-entry vehicles, but these teams were allowed to disband. Today, not many launcher designers really understand aeroplanes, and vice versa. Much of the analysis described here involves applying the techniques of airliner conceptual design to launch vehicles, but this approach went out of fashion several decades ago when the big aerospace companies stopped studying spaceplanes. I suspect that most aeroplane company design teams today, if tasked with working out a strategy for achieving the new space age soon and at affordable cost, would arrive at conclusions similar to those in this chapter.

A third explanation may be that introducing spaceplanes involves a complete re-conceptualization of our approach to spaceflight, and that large monopolistic government agencies are not usually among the prime movers of radical change. Three major changes will happen at more or less the same time. First is the change from launchers like missiles to those like aeroplanes, with the accompanying change in culture. Second is the change in government role from taking the lead to supporting and regulating the private sector. Third is the change in markets to dominance by large new commercial activities, especially tourism.

Whatever the explanation, if sensible evolution is suppressed for long enough, revolution follows. Progress will be rapid indeed as soon as the mindset changes.

So what is the way ahead? Richard Branson's Virgin Galactic is at present market leader in terms of developing a commercial spaceplane. They should be within three years of carrying the first fare-paying passenger. The chosen vehicle is *SpaceShipTwo* (SS2), which is an enlarged and improved version of SS1 shown earlier. Several other companies are working on suborbital spaceplanes for passenger carrying, and others are developing other 'new space' projects such as inflatable structures for space stations and mining asteroids. Most of these companies are based in the US.

It is very likely that some company will eventually make a commercial success of carrying passengers on brief flights to space. There is simply no reason why this should not happen. The technology is straightforward and there is great public interest in spaceflight. With economies of scale, maturing technology, and competition, the fare will come down from the present $100,000 to $200,000 to a few thousand dollars. There will be at least several dozen suborbital spaceplanes in service, each making several flights per day to space. The advantages of aeroplanes over missiles will then be clear for all to see, and the case for a fully orbital spaceplane will become unanswerable. One or more of these companies will probably then propose such a vehicle and a government space agency is likely to provide support. In this way, it seems likely that the missile mindset will be overcome and the development of the first orbital spaceplane will start within a decade or two. This will lead rapidly to a new space age, with greatly expanded space science and exploration, and with tourist visits to space hotels the largest commercial use.

This process could be greatly speeded up by planning for it now. The leading features of the first orbital spaceplane can be derived from a straightforward logic.

To achieve the new space age soon, we need to use existing technology as far as is practical. This requires the use of two stages. This is analogous to the in-flight refuelling used to extend the range of military aeroplanes. It adds complexity to a mission, but enables it to be achieved with existing technology. The amount of fuel needed for a flight to orbit using a single-stage vehicle with existing engines, measured as a fraction of take-off weight (about 87 per cent), is roughly equivalent to that needed to fly an aeroplane one and a half times round the world non-stop. The present record, held by the Virgin Galactic Voyager, is just once. To fly one and a half times around without in-flight refuelling would need either very advanced technology or the use of two stages – a large aeroplane carrying a specialised very long-range aeroplane part of the way. This analogy explains why flying to orbit with existing technology needs two stages. The lower stage boosts the upper stage to a speed such that it requires a practicable fuel fraction to carry on to orbit, and the two stages separate at this speed.

Single-stagers are clearly preferable in the long term, but these need very advanced new engines.

An airliner capable of flying to orbit would clearly transform spaceflight by providing vastly lower costs and improved safety. The essential design features of the first such vehicle can be derived from a straightforward design logic. The most important design requirement is greatly improved safety. To date, human spaceflight has achieved a safety level of about 100 launches per fatal accident. Imagine flying with an airline with such a safety record! Such poor safety is all but inevitable with throwaway launchers and compares with the more than one million flights per fatal accident achieved by scheduled airlines.

To achieve the required safety, the first orbital airliner should be as much like today's airliners as practicable, as these are the safest flying machines yet invented. It should therefore be piloted and have wings for taking off and landing using runways.

To achieve the new space age soon, we need to use existing technology as far as is practical. As discussed earlier, this requires the use of two stages.

So the spaceplane that will bring in the new space age will be designed for 'safety soon'. It will be piloted, have wings for conventional take off and landing, and will have two stages. Now, as mentioned earlier, many of the 1960s reusable launcher designs had precisely these design features. So the spaceplane that will revolutionise spaceflight will be similar to designs that could have been built in the 1970s!

This first orbital spaceplane could be built using only proven technology. Even so, it would be an ambitious project and there is a strong case for a less ambitious interim project. This could be smaller and suborbital only, but would retain the basic features of being piloted and having wings for takeoff and landing. Perhaps the most surprising feature of this situation is that the best demonstrator for this lead-in project that has actually flown was arguably the Saunders Roe SR.53 rocket fighter (Figure 11.6) that first flew in 1957! When it was cancelled as a fighter, Saunders Roe proposed a space research conversion, to be air-launched from a Valiant bomber and capable of suborbital flight. This proposal did generate considerable interest, but not enough to make it happen.

Fig 11.6: Saunders Roe SR.53 rocket fighter, which first flew in 1957. (*GKN Aerospace*)

It is interesting to speculate on what might have happened if the SR.53 had entered operational service as a fighter. Within a few years there would have been a reliable and mature aeroplane in service readily adaptable for suborbital flight to space. A two-seater derivative could have been built, and suborbital space tourism could have started in the 1960s! This would have led naturally to one of the orbital spaceplane projects being studied at the time and we would now be well into the new space age.

This saga is surely a supreme example of where we can and should learn from the history of technology.

Before turning to the more speculative prospects for re-engineering the human race, let us consider some 'sanity checks' that a proposed new technical development should be able to pass.

First, any realistic prediction should be in line with the laws of nature. Put another way, a good working assumption is that everything that exists is made up of matter and energy that follow natural laws. However, it should be remembered that our understanding of these laws changes over time. For example, if an inventor proposed a spaceship that could travel faster than the speed of light, the immediate objection would be that this is against the theory of relativity. The inventor might then point out that all the experimental evidence supporting this theory involves particles or bodies that are not self-propelled, in that their motion depends on forces of various kinds from other bodies, whereas his invention would indeed be self-propelled. So at least his superoptic spaceship could not be ruled out of hand as being theoretically impossible.

Similarly, an inventor proposing a device that appeared to break the second law of thermodynamics for example would have to show a plausible reason why our understanding of that law might change. However, travel in time does appear to be ruled out by this test.

A second test is that the inventor should be able to give at least a provisional answer to difficult questions. For example, someone proposing a driverless car for use on ordinary roads should be able to show how its sensors can reliably distinguish between a child and a dog, because most human drivers will take greater risks to avoid the child.

Further tests that an invention should pass if it is to be successful are that it meets a need, that it should be affordable, and that it can be made into a safe, reliable, and practical product.

A desirable test is that there should be no overriding moral or political objections to the development of the device in question. There are strong objections to nuclear weapons, for example, but probably a majority of people in the countries that have them consider that the security benefits outweigh these objections.

An invention that is an extrapolation of progress to date is generally more plausible than one that relies on breakthroughs.

A false test is that if a device is very difficult to imagine, it is not going to happen. Many developments that would once have been considered unimaginable have already taken place. If you went back in a time machine and tried to explain modern technology to Julius Caesar, for example, you might be able to give him some idea of how our mechanical devices worked; you might even begin to explain electrical gadgets; but there is no way you could even begin to explain the internet, mobile phone, or gene therapy. As Arthur C Clarke said in his 1973 revision of *Profiles of the Future*:'Any sufficiently advanced technology is indistinguishable from magic'!

Summarising the above, a good general rule in the history of engineering is that if something can be done and there is a reason for doing it, it eventually will be.

So, how do these tests affect the prospects of being able to re-engineer Homo sapiens using developments in computing and medicine? Our basic working assumption is that everything that exists is made up of matter and energy that follow natural laws. It follows from this assumption that the human body, mind, and soul can in principle be explained in scientific and engineering terms. Even our most powerful emotions must correspond to a particular pattern of cell activity in our brains. We 'just' have to find out what pattern of material building blocks results in human beings.

Put another way, finding out how we work is a question of reverse engineering. When an engineer wants to create a new product, he/she starts with an idea or a requirement. This is worked up into drawings and specifications, the design is analysed, and prototypes are built and tested until the device is ready for production. With reverse engineering, the engineer is presented with the finished product and then works back to drawings and specifications. If required, the device can then be put into production. Reverse engineering is used by industrial spies trying to find out how a superior rival product works, and in wartime to catch up with an enemy weapon that is better than one's own.

We have come some way towards reverse engineering the human body. At a very superficial level, there are mechanical, chemical, or electronic analogies to help to explain most of our parts. The heart, for example, is like a hydraulic pump; muscles are like electrochemical servos; the eye is like a video camera; the organs are like miniature chemical factories; the sub-conscious part of the brain is like a computer; and so on.

However, even in these grossly oversimplified terms, there is still one major conceptual mystery left – the nature of consciousness. If our basic assumption – that the total reality is made up of matter and energy – is correct, there must be a pattern of brain cells that enables us to actually 'feel' alive – to sense the outside world, to feel emotions, to fall in love, to appreciate music, and so on. At present, we simply do not know what this pattern is. There is no simple analogy yet available to us.

There may well be a conceptual breakthrough awaiting some great researcher that may in hindsight seem simple and obvious, analogous perhaps to the discovery of the double-helix structure of DNA. When is this eureka moment going to happen? If we extrapolate present growth trends then, by the late 2020s, the tool kit we use in artificial intelligence will include all the processes involved in human intelligence. (See, for example, 'Ray Kurzweil predicts the future', *New Scientist*, 21 November 2006). Kurzweil used the term 'The Singularity' to refer to what could happen when computers surpass human brainpower in all important respects, and this term is gaining ground. It may well take far longer than this – 2030 seems about the earliest for at least the clear prospect of the breakthrough. 'Immortalism' and 'Transhumanism' are other terms being used, but there has not yet been enough study of this subject for a reliable language to have evolved to describe it.

So, there appears to be a credible prospect of this fundamental breakthrough happening within a few decades. The nature of consciousness is certainly the subject of intensive research and progress is rapid. If and when this breakthrough happens, we will have a complete understanding of how the human body, mind, and soul work, at least at a conceptual level.

When this discovery is made, two outcomes seem possible. The first is that we find that, by having analysed precisely how we work, we will have destroyed our very humanity. We will have discovered that we are no more than computerised robots evolved in a universe that does not care whether we survive as a species or not. Our most sacred feelings were evolved merely to help us survive, and can be reduced to a (very large) table of numbers corresponding to brain activity. What happens next may not then seem to matter much. We may well blow ourselves up in a final orgy of self-indulgence.

The other outcome is that we find that we can re-engineer ourselves into a new life form, capable of even stronger love and spirituality, and feeling an even more powerful sense of oneness with the universe. Freed from the limitations imposed on us by evolution, we will be able to probe the fundamental capabilities of intelligent life.

Just as worms cannot imagine the emotions that monkeys can feel, and just as monkeys cannot imagine many human emotions, we cannot imagine what emotions and spirituality the 'post-humans' that we replace ourselves with will be capable of.

We can just about imagine a possible next step towards turning ourselves into post-humans. Computer chips, made of biological material, will be implanted in our brains. There will be direct computer to brain interaction, avoiding the need for screens and keyboards. As we learn more about how consciousness works, we will simulate it in the lab. Then we will be able to enhance our own consciousness, and communicate directly with others through an enhanced sort of 'telepathic internet'. We will be able to download our minds and souls into this internet. Then the Internet itself will become genuinely alive as a sort of prototype post-human, with all who want to becoming a part of it. The whole will be greater than the sum of the parts, and this first post-human will develop qualities and capabilities that we cannot yet imagine.

However, there is one important quality of this first post-human that can be predicted reliably – immortality. It need have no ageing or death genes and so can endure indefinitely. As hardware elements wear out, they can be replaced. The essential part of post-humans will be their memory, knowledge, emotions, and behaviour patterns, i.e., their software. This can be backed up and restored when necessary. Individuals who choose to join can have their own memory, knowledge, emotions, and behaviour patterns preserved indefinitely, which can be free to interact with the 'core' software as desired. In this way, these individuals will attain their own immortality, albeit as small cogs in a big machine. Those of us who are young enough may be able to witness what happens to this post-human, which may be able to live to the effective end of time.

What happens after this becomes increasingly speculative. A conceivable way ahead would be to start to influence the entire universe, perhaps building a sort of 'cosmic internet', until the limits of such post-human influence are reached.

The Singularity will certainly bring in some fundamental ethical challenges, involving the very basics of what it means to be human. This idea of humans being explainable in engineering terms will no doubt seem revolting to many. In a similar way, Darwin's claim that humans were no more than superior monkeys revolted many Victorians and led to a major crisis in church teaching. We got round this by a change in perspective. We came to regard monkeys as inferior humans. Likewise, we may come to appreciate that, if humans can be thought of as machines, machines can have human characteristics, even the ability to

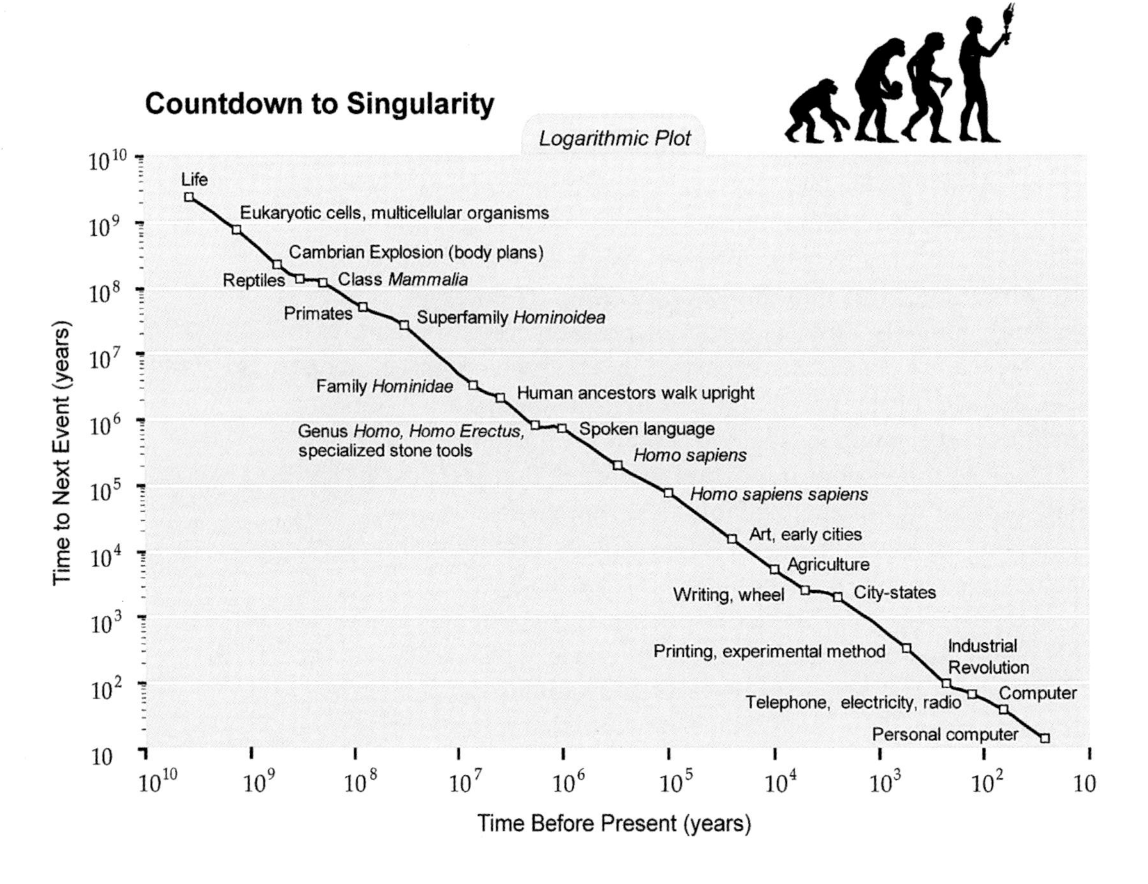

Fig 11.7: Technology development timeline on a log scale. (*Ray Kurtzweil*)

love. And post-humans will be not so much robots as enhanced humans. With this new perspective, the real miracle will be seen as the fact that a collection of matter and energy, assembled in a particular way, can feel alive.

The one thing that can be said for sure is that the rate of achieving major breakthroughs has accelerated exponentially over time. This is shown in Figure 11.7, taken from Ray Kurzweil.

Breakthroughs of very roughly comparable significance are plotted showing how long ago they happened. Approximately speaking, they fall on a straight line on a log scale, which indicates exponential development. The time to the next breakthrough is decreasing very rapidly. For example, moving from the industrial revolution to the personal computer happened some 1,000 times faster than from *Homo sapiens* to agriculture. Certainly, the runaway growth of electronic technology in recent years is without precedent. As mentioned earlier, Kurtzweil predicts that the next breakthrough will happen when computer power overtakes that of the human brain in all important aspects, and that this could happen within a few decades. He calls this turning point 'The Singularity', and there is already a Singularity University in Silicon Valley.

So, extrapolating the history of technology to date suggests the credible prospect that this century could well see the end of the human race as we know it. We will probably either self-destruct or go on to far better things.

Further reading

Ashford, David: *Space Exploration*, (Hodder and McGraw-Hill, 2013).

Clarke, Arthur C: *Profiles of the Future*, (Gateway, London, 1973 edn)

Dornberger, Walter: *The Rocket-propelled Commercial Airliner*, (University of Minnesota, 1956).

Kurtzweil, Ray: 'Ray Kurzweil predicts the future', *New Scientist*, 21 November 2006)

Kurtzweil, Ray: *The Singularity is Near*, (Viking 2006).

CHAPTER 12

The Heritage of Technology

Keith Falconer

Mankind has traditionally recognised and revered its technological achievements, whether by imagery such as rock art and cave paintings in prehistoric time celebrating hunting, or by detailed depictions of machines such as the siege engines on Trajan's column in Classical times or treadmill cranes in medieval manuscripts. From the Renaissance onwards the long-standing regard for collecting interesting technological artefacts – clocks, weapons, scientific instruments and models typified by the Medici Collection of Galileo's artefacts in Florence – led, by the nineteenth century, to the preservation of more mundane objects such as steam engines and machinery. In the twentieth century this was followed by the conservation of industrial sites and themed landscapes, and now in the twenty-first century to the recognition by World Heritage Site status of entire cultural landscapes based on technology and industry.

The origins of national concern for providing repositories or museums for objects of technological interest can be traced back to the mid-seventeenth century with the founding in Britain of the Royal Society, and in France of the Academie des Sciences, but in the eighteenth century, with many moves of premises in the case of the former and political vicissitudes in the case of the latter, little came in the way of permanent collections. During the French Revolution, in the initial enthusiasm for all things scientific, a centre had been established in Paris for the display of 'the wonders of engineering' which in 1794 was to become the Musée des Arts et Métiers which is still the national repository for the preservation of scientific instruments and inventions, and in 1793 a Military Balloon Research Centre had been set up at Meudon on the outskirts of Paris.

However, internationally, the main impetus was to be the Great Exhibition of 1851 in London. In its aftermath the South Kensington Museum was established celebrating industrial achievements, while the Patent Office Museum was formed to inspire further inventions. The two collections were to amalgamate in 1884 and the

Science Museum was born, housing Arkwright's spinning frame, an early Watt engine, Stephenson's Rocket and other iconic artefacts of the Industrial Revolution. Similarly, the Smithsonian Institution in Washington after the Centennial Exhibition of 1876 displayed American technology, while early in the twentieth century, the German Museum for the Masterpieces of Science and Technology was established in Munich in 1903, the National Technical Museum in Prague in 1908, and the Vienna Technical Museum, created in 1908, finally opened in 1918.

The first railway museums were opened at Hamar in Norway in 1896 and Nuremberg in Germany in 1899 and these inspired talk of doing the same in Britain, both in the 1890s and again in 1908 but this came to nothing at that time. The centenary of the opening of the Stockton & Darlington Railway in 1925 was marked by the opening in 1928 of a London & North Eastern Railway Museum in York. The smaller exhibits were housed in the old station buildings and appropriately the rolling stock and other large exhibits in the former locomotive erecting and repair shops of the old York & North Midland Railway. Numerous historic locomotives had of course been preserved elsewhere, including the very earliest such as *Puffing Billy* and *Rocket* in the Science Museum, *Wylam Dilly* in Edinburgh and *Locomotion* and *Derwent* in Darlington.

The preoccupation with preserving aspects of the history of technology greatly gained ground after the First World War with the focus widening to embrace sites. In Britain this led to the founding of the Newcomen Society for the Study of Engineering and Technology in 1920, to national surveys of historic bridges and wind and watermills, to the protection of notable industrial sites such as the Ironbridge as Ancient Monuments and to the preservation of half a dozen early steam engines *in situ* in Cornwall (see Figure 3.1). As a result of this early public nostalgia there are many hundreds of water and wind mills preserved in Britain and this is mirrored across the world, with international organisations sharing technical expertise (see Figures 1.3 and 1.4).

The Second World War witnessed the destruction of a great many historic industrial sites and the transformation of industry. But such losses and changes brought a new grassroots appreciation of heritage generally and industrial and technological heritage in particular.

In Britain, the threat of closure of canals in the 1950s and of railways in the 1960s led to popular protest campaigns, which resulted in the continued maintenance of most of the narrow-boat canal system for leisure and heritage (See Figure 6.6). Today there are more than 3,000 waterway sites designated as of historic significance, and technological advances in canal engineering in Britain at the end of the eighteenth century have been recognised by the inscription of the Pontcysllte Aqueduct and its landscape as a World Heritage Site (see Figure 6.4). Similarly, later waterway engineering feats such as the Anderton Boat Lift of 1875 have attracted state funding for restoration and even funding for a modern technological wonder, the Falkirk Wheel, which re-unites the Forth & Clyde Canal with the Union Canal some 82 feet (25m) above. Opened in 2002, it attracts some 400,000 visitors per annum.

Similarly, the British affection for railways in the face of the Beeching cuts of the 1960s led to the preservation of numerous small railway lines on which over 1,000 steam locomotives are operated by enthusiast societies. Indeed, following public outcry at the demolition of Euston Station and its famous Arch in 1962 there was to be a complete reversal of attitude by the national bodies managing the railways and now many historic stations have been sensitively modernised and restored, and the Forth Bridge, that icon of pioneer steel construction, is now a World Heritage Site.

Fig 12.1: The Falkirk Wheel, Scotland. (*Author*)

The public desire to celebrate technological maritime achievements was soon to follow. Famous ships such as HMS *Victory* and the *Cutty Sark* had long been preserved for their historic interest and have recently been joined by the remains of that marvel of Tudor technology – the *Mary Rose,* (see Fig 5.3), but now the technological interest of vessels such as Brunel's SS *Great Britain* (see Fig 6.7 & 6.8) and the ironclad HMS *Warrior* (see Fig 5.5) captured public attention which also extended to twentieth century warships such as the HMS *Belfast* and also to submarines. There is now a new museum at Gosport, near Portsmouth, displaying the development of British submarines while, at the Cite de la Mer naval museum at Cherbourg, *Redoutable*, a 1971 nuclear submarine, has been preserved with the reactor compartment replaced by a new section. She is the only complete ballistic missile submarine open to the public.

Global interest in the age of the ocean liners led, on the retirement of RMS *Queen Mary*, to her being preserved as an hotel and tourist attraction at Long Beach California and to other museums celebrating notable vessels such as the ill-fated *Vasa* and the *Titanic*. The naval dockyards that built and supported the fleets of warships and the ports that served worldwide maritime trade have also been preserved and some such as the Royal Naval Dockyards of Chatham and Portsmouth are now popular public attractions, while Liverpool Docks are part of a maritime themed World Heritage Site.

The 1970s and 1980s also saw the creation of hundreds of small preservation trusts and numerous local authority museums dedicated to the preservation, display and in many cases operation of historic machinery and processes. Prominent amongst these were water and wind driven mills and steam engines. The steam engine in all forms – stationary and moveable – has attracted a great deal of attention. Indeed over 100 stationary steam engine sites have been preserved *in situ* in Britain alone, representing over 150 years of steam engine evolution, while countless more locomotives and steam powered vehicles are preserved in working condition.

Amongst the technological sites of great significance illustrating the development of the stationary steam engine is the sole surviving *in situ* Newcomen Engine dating from 1795 at Elsecar Colliery, the earliest working Cornish steam engines which supplied water to the Kennet & Avon Canal at Crofton and the magnificent collection of early steam engines at Kew Bridge Pumping Station. This fascination with steam is mirrored across the world and indeed

Fig 12.2: The *Vasa*, sunk on maiden voyage in 1628 and raised in 1961, on display in Stockholm. (*Peter Isotalo*)

Fig 12.3: 1812 Beam engine at Crofton Pumping Station. (*Wikipedia, Chris J Wood*)

more than 1,500 websites are devoted to preserved stationary steam engines.

While mills and steam engines, and especially those capable of still being operated, may have captured the early limelight, attention widened to preserving mining, processing and manufacturing sites and to museums dedicated to industry. In Europe in recent years an umbrella organisation, the European Route of Industrial Heritage (ERIH), has emerged from a small European Union funded regional pilot project to span the continent. ERIH is now an independent entity funded by its members and is proving outstandingly successful, it has expanded greatly geographically and its membership is much sought after in countries with less-developed industrial heritage infrastructure. It currently presents more than 1,000 sites of industrial or technological interest in forty-four countries with seventy-seven Anchor Points, seventeen Regional Routes and thirteen Theme Routes including Textile Mills, Iron and Steel, Mining, Energy, Papermaking and Salt Production. In the United Kingdom since 1994 the Heritage Lottery Fund (HLF) has been a major funding stream for the preservation of technological heritage. In the industrial, maritime and transport sector it has funded over 3,000 projects totalling over £1 billion of HLF money, which has levered in twice that in matching funds.

Towards the end of the twentieth century a major leap forward in the recognition of technological virtuosity was the inscription by UNESCO of industrial sites as World Heritage Sites. Many of these were single sites such as Engelsberg Ironworks in Sweden, the Royal Saltworks of Arc-et-Senans and the Canal du Midi in France, but in 1986 thematic industrial landscapes were pioneered by the UK with the inscription of the Ironbridge Gorge World Heritage Site embracing not only the famous bridge (see Fig 6.2) but the relic canal features and nearby furnaces, engineering works and settlements (see Fig 3.3). This concept was further elaborated in the UK Tentative List that was published in 1999 which led, in Britain, to the inscription of other thematic landscapes such as the textile mills of the Derwent Valley, Saltaire and New Lanark (see Figs 3.3 and 10.2), Blaenavon Industrial Landscape (see Fig 3.5), Liverpool Maritime City and the Cornish mining landscapes (see Fig 3.1) and prompted the inscription of similar themed landscapes across the world, including the Tomioka Silk Landscape in Japan, saltpetre landscapes in Chile and the Major Mining Sites of Wallonia where the sites form a strip 106 miles (170km) long by 2–9.3 miles (3–15km) wide, crossing Belgium from east to west.

The 2012 inscription of Nord-Pas de Calais Mining Basin as a World Heritage Site took the concept even further by having the ambition to preserve cultural traditions as well as tangible remains. Remarkable as a landscape shaped over three centuries of coal extraction, the site's 109 separate components feature pits, slag heaps, transport infrastructure and mining towns and villages. The site bears testimony to the quest to create model workers' cities from the mid nineteenth century to the 1960s and documents the living conditions of workers and the solidarity to which it gave rise. It is also very much a monument to, and product of, technological advances in mining – much of the landscape would be inundated by a rise in the water-table if pumps were not continuously employed.

A similar aspiration was the inscription of the town and meat processing factories of Fray Bentos in Uruguay as a World Heritage Site. Here immigrant workers from more than forty different countries were, in the course of a century, to create on a greenfield

Fig 12.4: Nord-Pas de Calais Mining Basin, The pithead buildings at Loos-en-Gohelle. (*Wikipedia, Jérémy-Günther-Heinz Jähnick*)

site on the River Uruguay a remarkable landscape manifestation of the globalised food industry.

A particular challenge for the technological heritage is to keep evidence of changes over different periods. The staple industries of the Great Age of Industry – coal, iron and steel, textiles and heavy engineering – peaked by the middle of twentieth century and have largely disappeared in many countries. They have been replaced by twentieth century creations such as the car, aircraft, space and electronic industries, the service and leisure industries and the food and beverage industry, and these industries have also undergone great changes and are equally part of technological heritage. They too have been recognised as World Heritage Sites, with the Fagus Factory in Alfeld, the modern Zollverein Coal Mine and Van Nellefabriek complex in the Netherlands all inscribed as World Heritage Sites in recent years.

There are now numerous sites and museums dedicated to these later industries and prominent amongst these are car museums and aviation and space museums. There are many thousands of vehicles preserved around the world, with more than 4,500 in the top twenty car museums in the United States, while the company museums of Renault and Peugeot in France have more than 1,200 cars preserved between them and in Britain the three largest national car museums have some 1,000 cars on display. Most of the car museums are modern custom built structures but the Ford Piquette Avenue Plant in Detroit is a museum in the former mill type car factory built in 1904 where the Ford Model T was first developed and built in 1908. In January 1910, after assembling nearly 12,000 Model Ts at the Piquette Avenue Plant, Henry Ford moved production to his Highland Park complex where he introduced the moving assembly line 1913–1914 and would eventually produce 15 million Model T Fords.

Wikipedia lists more than 600 aviation and space museums, of which some 250 are in the United States. Some of these sites are of some antiquity even predating powered flight – the Chalais-Meudon near Paris for example developed out of the Military Balloon Establishment created in 1877. The huge wind tunnel constructed 1932–4 to test aircraft with a wing span of up to 39 feet (12m) was in the 1970s used to test the scale model of Concorde. The National Air and Space Museum of the Smithsonian Institution holds the largest collection of historic aircraft and spacecraft in the world (see Fig 11.4). It was established in 1946, as the National Air Museum,

Fig 12.5: Ford Piquette Plant, (1904), Detroit. (*Wikipedia Rsepo9*)

and opened its main building in 1976. In 2014, the museum saw approximately 6.7 million visitors, making it the fifth most visited museum in the world. The US Air Force Space & Missile Museum located at Florida is also of note and includes artefacts from the early American space program and an outdoor rocket garden displaying rockets, missiles and space-related equipment chronicling the US Air Force.

In Britain the Science Museum Group, consisting of the Science Museum in London, Museum of Science and Industry in Manchester, National Railway Museum (York), National Media Museum in Bradford and National Railway Museum (Shildon), is devoted to the history and contemporary practice of science, medicine, technology, industry and media. With five million visitors each year and an unrivalled collection, it claims to be is the most significant group of museums of science and innovation worldwide.

In specialised fields such as medicine, there has long been an appreciation of exhibits as a teaching resource. The Royal College of Surgeons of Edinburgh, for example, was founded in 1505; with the

Fig 12.6: Ballistic missiles in the National Air and Space Museum, Washington DC. (*WikipediaTurelio*)

Museum's collections growing from 1699 after 'natural and artificial curiosities' were publically sought. In the 1800s, the Museum had expanded to include the remarkable collections of Sir Charles Bell and John Barclay. Preserving much more recent developments, the Thrackray Medical Museum which opened in 1997 in Leeds enables the public to learn the story of medicine and is housed in a former Victorian workhouse that was latterly a NHS hospital. Similarly the Wellcome Collection founded in 2007 in London is a museum displaying an unusual mixture of medical artefacts and original artworks exploring 'ideas about the connections between medicine, life and art' and attracts over 500,000 visitors per year. The Science Museum's gallery The Science and Art of Medicine is one of the greatest collections about the history of medicine in the world. Told through a display of 5,000 objects the gallery is currently being greatly expanded.

The technology of weaponry and explosives is equally celebrated by both old and recent museums and preserved sites around the world. In Britain, the Royal Armouries, in addition to its traditional home in the Tower of London, have important collections housed in a new museum in Leeds, and at Fort Nelson, a vast Victorian fortification built in the 1860s as part of a defensive chain around Portsmouth and its vital Royal Dockyard. Fort Nelson is home to the Royal Armouries' collection of artillery and historic cannon – the Big Guns – with over 350 on display, including the massive 200-tonne railway gun. Nearby, the Museum of Naval Firepower is housed in the historic setting of a former gunpowder and munitions depot at Priddy's Hard, on the Gosport side of Portsmouth Harbour.

At Waltham Abbey in Essex the Royal Gunpowder Mills were established in 1787 on an existing gunpowder site and developed over the next two centuries as the principal government research and production site for explosives of all sorts. Decommissioned in 1991 it opened as a public attraction in 2000 set in 175 acres of parkland and woods containing twenty-one buildings of major historical importance served by an internal canal system on two levels.

Similar sites have been preserved across the world. For more than 200 years, from its foundation in 1758, the Gunpowder Works in Frederiksværk was the principal supplier of gunpowder and explosives to the Danish armed forces, and the open-air museum offers a glimpse into the history of the manufacture of explosives as well as the process of industrialisation in Denmark. In the US the Hagley Museum is located on 235 acres along the banks of the Brandywine in Wilmington, Delaware, and is the site of the gunpowder works founded by E I du Pont in 1802. This example of early American private industry includes restored mills with nineteenth century machinery, a workers' community, and the ancestral home and gardens of the du Pont family.

This brief introduction to preserved technological artefacts and sites shows that the appreciation of the significance of the heritage of technology has come a long way over the last three centuries since the tentative interest of a few learned

societies and rich patrons. It is now a worldwide activity, part-funded by governments, and embracing thousands of sites and museums with income from millions of visitors. With its own specialised international networks of professionals, codes of practice, sophisticated conservation ethics and techniques, and ever-changing ambitions, the heritage of technology has become an industry and technology in its own right.

Fig 12.7: Royal Gunpowder Mills, Waltham Abbey, steam driven incorporating mills. (*Geograph, Christine Mathews*)

About the Authors

David Ashford

David Ashford is Managing Director of Bristol Spaceplanes Limited, an innovative small company developing the *Ascender* spaceplane. He graduated from Imperial College in aeronautical engineering and spent one year at Princeton doing post-graduate research on rocket motors. His first job, starting in 1961, was with the Hawker Siddeley Aviation spaceplane design team, working on spaceplanes among other projects. He has since worked as an aerodynamicist, project engineer and project manager on various aerospace projects, including the DC-8, DC-10, *Concorde,* the Skylark sounding rocket, and various naval missile and electronic warfare systems at Douglas Aircraft and at what is now BAE Systems. He co-authored with Prof. Patrick Collins the first serious book on space tourism, *Your Spaceflight Manual: How You Could be a Tourist in Space Within Twenty Years* (Headline, 1990), and wrote a follow-up book, *Spaceflight Revolution* (Imperial College Press, 2002). His latest book is *Space Exploration: All That Matters* (Hodder, (2013). He has had published about twenty papers on space transportation in the professional press.

Mike Bone

Mike Bone studied history, education and management at the Universities of Wales, Reading and Bath. Most of his career was spent in post-compulsory education, finishing up at a government agency where he specialised in corporate governance, strategic management, quality and standards, taught on an MBA programme and assessed NVQs in management.

He has been active in the study and conservation of industrial heritage since the early 1970s, is a past chair of the Association for Industrial Archaeology, the Bristol Industrial Archaeological Society, and the Avon Industrial Buildings Trust Ltd, and a member of advisory panels/advocacy groups of the Heritage Alliance, the Heritage Lottery Fund and English Heritage. He is currently chair of Bristol Industrial Archaeological Society and the Avon Industrial Buildings Trust Ltd, a member of the committee of the Brewery History Society and Bristol City Council's Conservation Advisory Panel and a member of the History of Technology Research Unit (HOTRU) at the University of Bath, where he is a visiting research and former Rolt fellow.

His publications on industrial archaeology include work on Devon, Dorset, Leicestershire and, latterly, the Bristol and Bath area. He was a convenor and contributor (with David Dawson) for the post-medieval, industrial and modern periods of English Heritage's South West Archaeological Research Framework. Recent publications have included commentaries for The Godfrey Edition of reprinted Ordnance Survey maps of the Bristol area and (with Tim Edgell), *Brewing in Dorset* (2016).

Angus Buchanan

Angus Buchanan is Emeritus Professor of the History of Technology at the University of Bath. He was born in Sheffield in 1930 and went to High Storrs Grammar School there. Then he went to St Catharine's College Cambridge, where he took a B.A. in History, and subsequently went on to take an M.A. and a Ph.D. He has been the Director of the Centre for the History of Technology at the University of Bath since its inception in 1964. He has served terms as President of the Bristol Industrial Archaeology Society, the Association for Industrial Archaeology, the Newcomen Society for the History of Engineering and Technology, and of the International Committee for the History of Technology. Professor Buchanan has also served as a Visiting Lecturer at the University of Delaware, U.S.A.; and as a Visiting Fellow at the Australian National University, Canberra; a Visiting Lecturer at the Technological University of Wuhan, China; and as a Visiting Professor at Chalmer's University, Gothenburg, Sweden. His publications include: *Industrial Archaeology in Britain* (1972); *The Power of the Machine* (1992); *The Engineers: A History of the Engineering Profession in Britain* (1989); and *Brunel: The Life and Times of I K Brunel* (2002). Professor Buchanan was appointed OBE in 1992 for services to the History of Technology.

Brenda Buchanan

Brenda J Buchanan B.Sc. (Econ) Ph.D. (Univ. of London) FSA Dr Buchanan has been a Visiting Research Fellow at the History of Technology Research Unit of the University of Bath since 1987.

Her main research interest concerns the history and technology of gunpowder. She has edited and contributed to *Gunpowder: The History of an International Technology* (Bath University Press, 1996, reprinted 2006) and *Gunpowder, Explosives and the State: A Technological History* (Ashgate Press, Aldershot, 2006). She has also made invited contributions to Brett D Steele & Tamera Dorland eds: *The Heirs of Archimedes. Science and the Art of War through the Age of Enlightenment* (MIT Press, USA, 2005) and Brenda Buchanan et al, *Gunpowder Plots* (Penguin Books, 2005). Of the three articles on this theme published in *ICON Journal of the International Committee for the History of Technology,* the most recent was in volume 20 (2014). It was entitled 'Gunpowder Studies at ICOHTEC'. After many years helping to explore and establish the subject at the national and international level she now focuses on her own research.

Keith Falconer

Keith Falconer has been involved with industrial heritage since the late 1960s. An MA at Edinburgh University was followed by three years research on industrial relic landscapes at Hull University. In 1971 he was appointed CBA Industrial Monuments Survey Officer tasked with identifying for protection historic industrial sites throughout the UK. Based at Centre for the Study of the History of Technology at the University of Bath the post transferred in 1981 to the Royal Commission on the Historic Monuments of England as Head of Industrial Archaeology. On merger with English Heritage in 1999 he continued in this role until he retired in 2012.

Author of *Guide to England's Industrial Heritage* and *Swindon: The Legacy of a Railway Town* he has written numerous articles on the management of industrial heritage. Since 1998 he was been involved in developing many industrial World Heritage Sites and advising the Council of Europe on European industrial heritage. He is a member of the Canal & River Trust's Heritage Advisory Group, a past Chairman of the Association for Industrial Archaeology and a Visiting Research Fellow at the University of Bath He was awarded an OBE for services to industrial heritage in 2013.

Richard Harvey

Richard Harvey is an Emeritus Consultant Physician in Bristol. He trained at the Middlesex Hospital Medical School, London. After a series of posts in clinical medicine, he worked at the Medical Research Council Gastroenterology Unit at the Central Middlesex

Hospital and then at the Institute of Nuclear Medicine, Middlesex Hospital. In 1971 he moved to Bristol to work in the Academic Unit of Medicine in the University of Bristol, based at Bristol Royal Infirmary.

In 1976, he became Consultant Physician to the Gastroenterology Unit at Frenchay Hospital, Bristol, and Clinical Senior Lecturer in Medicine at the University of Bristol, where he continued with a combination of clinical work, teaching and research. He has written over 100 scientific papers and two books, *Clinical Gastroenterology and Hepatology* and *Basic Gastroenterology*, both with Professor A E Read.

Stephen K Jones

Stephen Jones is the Member for Wales and Chainbridge Sub-Panel Convenor on the Institution of Civil Engineers (ICE) Panel for Historical Engineering Works, and is a Companion member of the ICE.

His background is in economic development, latterly with the Welsh Development Agency, where he specialised in new technology and innovation programmes. Following a secondment with ICE Wales Cymru and Cadw, he now acts as an engineering heritage consultant. Industrial and engineering history has been a long term interest and Steve enjoys writing and speaking on the subject, especially the works of Brunel. In 2005 the first of his trilogy on *Brunel in South Wales* was published, and this was completed in 2009. Steve is a visiting lecturer at Swansea University and an Associate of the History of Technology Research Unit at Bath University.

Robin Morris

Dr Peter Robin Morris spent a long career in the electronics industry as a semiconductor engineer in both British and American firms, before taking the position of Senior Lecturer in the Department of Systems and Communications Engineering at the Southampton Institute of Higher Education. Dr Morris received a Diploma in Communications Engineering and Electronics from Southampton Institute in 1962 and was elected C.Eng in 1968. He later received a B.A. degree in History from the Open University followed by an M.Phil. from the University of Bath, and then a Ph.D. from the Open University. He was subsequently appointed to a Visiting Fellowship as a Rolt Fellow at the University of Bath. In 1990 he published *A History of the*

World Semiconductor Industry (Peter Peregrinus Ltd), volume 12 in a series prepared by the Institution of Electrical Engineers. He is now retired and lives in Malvern.

Giles Richardson

Giles Richardson is a doctoral research student based at the Oxford Centre for Maritime Archaeology. He graduated from Durham University in Ancient History and Archaeology and completed an M.A. in Maritime Archaeology at the University of Southampton. His interests include the archaeology of naval warfare and ancient seafaring technologies. He has worked across Europe and the Middle East as a field archaeologist, including spending a year at the British School in Rome. He is currently part of an Anglo–French mission excavating at the sunken Egyptian city of Heracleion - Thonis, and regularly surveys shipwrecks around the UK coastline.

Owen Ward

Owen Ward is a long-standing member of the Bristol Industrial Archaeological Society and of the International Molinological Society. He retired in 1990 from an administrative position in the School of Humanities and Social Sciences at the University of Bath and was appointed as a Visiting Research Fellow in the History of Technology Research Unit. He has taken an interest in the history of the production of millstones, especially the French Burr stones manufactured in or distributed from La Ferté-sous-Jouarre, and has also studied the history of local milling industries such as paper and early textile manufacture.

Index